19·60

Alexander Heinz
Folding Polyhedra

Text and photos, unless otherwise stated:
Alexander Heinz, DE-Herdecke, geomenta.com
Design and typesetting: Frank Georgy (Kopfsprung.de), DE-Cologne
with the assistance of Alexander Heinz
Illustrations: Alexander Heinz

Originally published as *FALTPOLYEDER: Papierfalten zwischen Kunst und Geometrie* by Haupt Verlag
Bern, Switzerland © 2019

Translated from the German by Katrin Binder, assisted by Neil Franklin and Alan Stott.

Library of Congress Control Number: 2020943517

Edited by Ian Robertson

Type set in Brandon Grotesque
ISBN: 978-0-7643-6157-9
Printed in China
Published by Schiffer Publishing, Ltd.
4880 Lower Valley Road
Atglen, PA 19310
Phone: (610) 593-1777; Fax: (610) 593-2002
E-mail: Info@schifferbooks.com
Web: www.schifferbooks.com

For our complete selection of fine books on this and related subjects, please visit our website at www.schifferbooks.com. You may also write for a free catalog.

Schiffer Publishing's titles are available at special discounts for bulk purchases for sales promotions or premiums. Special editions, including personalized covers, corporate imprints, and excerpts, can be created in large quantities for special needs. For more information, contact the publisher.

We are always looking for people to write books on new and related subjects. If you have an idea for a book, please contact us at proposals@schifferbooks.com.

SCHIFFER PUBLISHING

4880 Lower Valley Road • Atglen, PA 19310

GINKGO BILOBA

In my garden's care and favour
From the East this tree's leaf shows
Secret sense for us to savour
And uplifts the one who knows.

Is it but one being single
Which as same itself divides?
Are there two which choose to mingle
So that each as one now hides?

As the answer to such question
I have found a sense that's true:
Is it not my songs' suggestion
That I'm one and also two?

JOHANN WOLFGANG V. GOETHE

(TRANSLATION BY JOHN WHALEY)

Alexander Heinz

FOLDING POLYHEDRA

THE ART & GEOMETRY OF PAPER FOLDING

SCHIFFER PUBLISHING
4880 Lower Valley Road • Atglen, PA 19310

Contents

Contents

Foreword

For years, participation in the International Conference of Geometry in Strobl, on Lake Wolfgang, and the presentation of a paper there have been among my fixed engagements as professor in geometry at the University of Applied Arts Vienna. This conference is a central information-and-communication hub for all those teaching or studying in the field of geometry. There, one is sure to meet familiar faces and many speakers who do not mind—I will freely admit it—the somewhat strenuous journey into the stunningly beautiful lakeland area on the border between Salzburg and upper Austria, to be carried away afresh each time by the palpable enthusiasm for everything related to geometry. One of these familiar faces is Alexander Heinz, who regularly speaks at the conference, presenting and exhibiting his paper polyhedra and leading workshops on how to obtain the desired results.

On a bus journey from Salzburg to Strobl—surrounded by suitcases and parcels containing his models—he once told me that he had received the Phänomena prize of the German Society for Geometry and Graphics (DGfGG) for the most beautiful model for his polyhedra, but really for his entire way of working. And he said to me: "You find yourself at an interesting junction between mathematics and art. From this perspective you have a special approach to geometry, and you gain much experience in the interaction between these two very different departments. Art for me is the urge to create something new, and to explore it to understand it. Art and science, from that point of view, are two sides of the same coin." "Yes," I said, "I completely agree! To understand geometry as the union of science, art, and craft is a beautiful endeavor. Too long have the fields of art and geometry been separated after their wonderful coexistence during the Renaissance period."

Very early findings from the British Isles bear witness to the fact that polyhedra forms have accompanied human beings for several thousand years. From ever-changing forms of presentation, conclusions can be drawn regarding the awareness of space of the respective period. Alexander Heinz's folded models show clearly that we can still discover new aspects in such forms. The most-divergent aspects can be brought together in the geometry of polyhedra. It plays a prominent role in chemistry (in the construction of molecular structures), in crystallography (for the classification of crystal shapes), and of course in geometry. Some pathogens occur in the form of icosahedra or dodecahedra, and all ball shapes and designs are ultimately based on platonic solids. This is evident in the prehistoric carved-stone balls from the British Isles, which effectively count as the precursors of the ball shapes we use today.

Even the design of play structures (such as climbing frames) and the architecture of roof construction or domed buildings, as well as radar domes, derive from polyhedra shapes. Polyhedra offer spatial pictures of mathematical ratios, and their own aesthetic use of forms is understood even outside the scientific community. Alexander Heinz's book is an example of how art, crafts, and mathematics can be united in a simple and at the same time universal way. It meets the human fascination with astonishing mathematics while presupposing hardly any prior knowledge. Eventually, Alexander Heinz's polyhedra are simply regular shapes that are folded symmetrically and then put together. This, however, is not at all trivial: starting from the simplest observations, such as the counting of vertices, edges, and surfaces, reveals hidden numerical spatial references that deepen our abilities to think and act with more spatial awareness. We are used to conceiving of polyhedra as made up of surfaces. With the models in this book, we have to start our observations from the vertices. This requires constant, active mental participation. In this way, Alexander Heinz's book also invites us to reconsider and reshape familiar thought patterns.

Folding Polyhedra is a delight because of its appealing layout alone, yet the reason why this book should not be missing from any library is its wealth of detail and its systematic treatment of the subject. According to tradition, only those allowed to enter Plato's Academy (the philosophers' school founded by Plato in ancient Athens) were well versed in geometry. It was reckoned as the foundation for every academic discipline. And in the Middle Ages, geometry was taught as one of the seven liberal arts (Lat.: artes liberales). It was thought that the education of future regents had to include a knowledge of geometry. Those who intended to rule over a people had to train their thinking and behavior to be as precise as possible in spatial terms. Today, any one of us is free to engage with geometry. In this sense it may be hoped that Alexander Heinz's book will find many readers who take pleasure in constructing paper models and who enjoy geometry for its own sake.

Georg Glaeser
Professor in mathematics and geometry at the University of Applied Arts Vienna

Folded Polyhedra: A Union of West and East

Two different subjects—geometry and origami—have been combined into one in this book: more precisely, regular and semiregular polyhedra shapes, with the building of simple models using paper. Polyhedra (Engl.: many-faced shapes, or shapes with many surfaces) and their geometry can be traced 5,000 years in Western cultural tradition. Their realization in Folding Polyhedra is achieved with the simplest methods of folding and constructing, using paper as our material (origami), a technique that originated in an Eastern cultural tradition with an antiquity of some 3,000 years. In this book, both mutually benefit from each other. There always remains something to be discovered in the age-old geometric forms, and equally the techniques of origami can be continually developed further.

Modular Structure
Initially, the step from the second dimension (the plane) to the third (spatial) dimension presents a barrier for the inexperienced imagination. The origami method allows us to scale this barrier in a playful way. The finished model of a polyhedron is spatially and aesthetically more than the sum of its individual parts. All models in Folding Polyhedra are realized in the same way: the center of the individual paper base becomes the crossing point (see p. 9, figure on the left) through symmetral (symmetrical) folding. This crossing is the center point of each respective module, consisting of two differently folded basic

units (see horse and rider, p. 30). Each polyhedric model is built from several such basic units. In this book, polyhedra are exclusively formed from regular polygons (from equilateral triangles, squares, and pentagons, as well as hexagons, octagons, and decagons). This modular method becomes easily accessible in the construction process of the first model (see also p. 38), an octahedron: the edges of the polyhedron to be built are formed when the individual (vertex) modules are joined (see p. 9, figure on the right). This process is repeated until all modules are connected with each other through their edges. The edges join in a ring (through a ring joint) to form polygons that together make up the polyhedron. For the octahedron, we start with a square base that is folded symmetrically into what is called "horse and rider" modules (see p. 30) and then joined to form triangular ring joints (see p. 9, figure on the right). The result is an octahedron formed by eight triangular sides.

From Base to Model
In the majority of cases, the surface shape of the paper base that we start from and the shape of the faces of the polyhedron that are eventually seen in the finished model will not be identical. When talking about "surface," we consequently face the necessity of distinguishing between the respective shape of the paper base we start from and the surface that is formed by the ring joint of the corner modules in the polyhedron model. For a

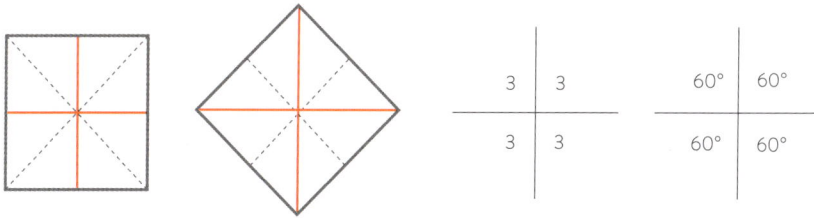

Schematic illustration of the base surface (a horse-and-rider module) of the octahedron model

Each octahedral module forms a four-edged vertex of the model. Here, four triangles (3) with 60° come together (schematic illustration).

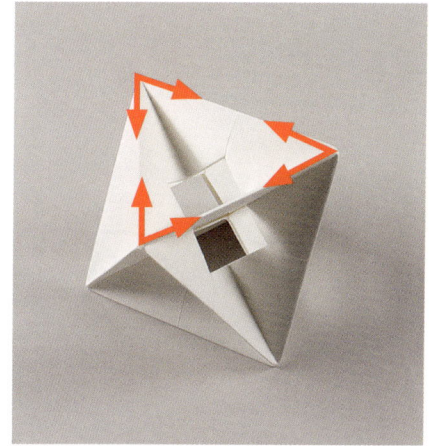

Several horse-and-rider modules are joined to form closed (angular) ring-shaped joints.

cube model (see p. 76), for example, a quadrinomial ring joint (a ring-shaped joint across four edges) is needed to form the square-shaped surfaces of the cube. For the vertices, where three edges meet, the principle followed in this book demands folding pieces of paper that are triangular in shape. The relation between the shape of the folding paper and the form of the ring joint reveals—as we will see later—basic numerical relationships in the polyhedra. Each polyhedron has a dual partner, something that is illustrated well by the example of the cube and the octahedron: A square has eight corners and six surfaces, while in an octahedron it is the exact opposite. In view of this fact, the construction of polyhedra models from corner modules receives a special dimension of interest and a deeper meaning.

Regular and Semiregular Folded Polyhedra

The octahedron was the point of departure of a course of development over many, many years, the incentive to which and inherent aim was to realize all regular and semiregular polyhedra using the same technique. The names of the individual models refer to the particular polyhedra forms that emerge. The results are realized in a series of white models, all of which have the same size or (as closed polyhedra shapes) the same volume. The sizes given for the individual models (see table, p. 32) represent approximate values, allowing for a realization of the models without having to make tedious calculations. The basic shapes

for each model have been individually adjusted accordingly. The last polyhedra of the white-model series require much patience, delicate fine motor skills, and experience. To start with, it is advisable to follow the given sequence of individual folded polyhedra. It is best if you begin with the simple shapes and postpone tackling the more difficult models until you have built up some experience. To facilitate the beginning, the white-model series has been supplemented with step-by-step instructions for colored models. A comprehensive list of important hints for practical work is found in "Practical Considerations," starting on page 22.

Extended Polyhedra Forms

The forms in chapters J, K, L, and M (see p. 120 and following) are simpler than the majority of the demanding regular and semiregular models. Here, some of the regular and semiregular forms are modified and extended. These new forms follow the same construction principles. Beyond these, other and freer forms are possible, but instructions for these are reserved for separate description in a future publication.

Historical Background

Platonic solids (see the glossary, p. 176) are, like all elements of geometry, initially nothing but ideas. Without an actual, material model that illustrates this idea, it is impossible to specify the size of, for example, a cube, the thickness of its walls, and whether it is solid or hollow. Merely the shape of its faces, the angle in which the edges meet at the vertices, and how many faces, edges, and vertices it has can be described. Over the centuries there have been numerous divergent illustrations of cubes. Each emphasizes certain aspects, which in turn reveal something of the concept of space in the period in question.

Carved-Stone Balls
Carved-stone balls (see p. 11, figure on the left) are prehistoric stone spheres about the size of tennis balls that have been found in the British Isles. Their age can be dated only approximately to about 2,000 to 3,000 years BCE. A large part of the about 400 specimens from a range of different types of rock were found near Aberdeen, Scotland. Outside Scotland, further specimens were found in England, Ireland, Wales, and Brittany, all regions that are known for their prehistoric stone settings (menhirs). Carved-stone balls exist in different forms, and their workmanship is not uniformly fine. Apart from many other forms, there are stone balls that correspond to the symmetries of all five platonic solids. They could also be called "spherically rounded solids," which are similar to the more fully

differentiated ball types used in today's sports. The prehistoric carved-stone balls still pose a riddle for scholars; their use still remains a mystery. Even though these forms were worked very thoroughly in many cases, little imprecisions show that they were made by hand. In detail, their execution is not quite as exact as contemporaneous masonry from Egypt.
Frank Teichmann (1937–2006), an expert in Egyptology and archeology, as well as author of several books concerning a range of cultural epochs, has called the megalithic culture a "culture of movement." This term takes into account the structure of the landscape, climate, and agricultural methods of the individual regions, but also the astronomical orientation of the stone placements. The spherical form of the carved-stone balls also seems to reflect this: if geometrical solids are to roll, they have to be transferred to spherical forms.

Flower of Life
Nearly at the same time as the development of the carved-stone balls, the richly documented Egyptian civilization finds its fullest expression. In contrast with the Stone Age civilization of western Europe, this is characterized by harmonious proportions, immutability, exactness, and perfection. Initiates hermetically screened their knowledge from the uninitiated public. The plane or wall is in many ways characteristic of this civilization. There is an illustration of Platonic solids corresponding to

Carved Stone Balls

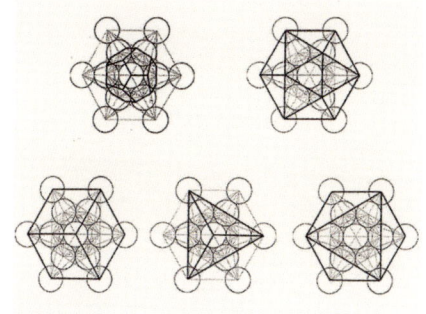

Flower of life / Metatron

Regular polyhedra following Plato's description

the seemingly static, still inherently two-dimensional character of the Egyptian civilization standing at the threshold to three-dimensionality: six circles gather around a central circle. It is possible to project any Platonic solid into this pattern by drawing a line between the centers of the circles or other prominent points of construction. In esoteric circles, this image is a symbol of cosmic order. It is said to have energizing, harmonizing, and protective effects. As the so-called flower of life (see figure on the right) or Metatron cube (compare with bibliography, Drunvalo Melchizedek, p. 178), it is vaguely ascribed to Egyptian sources. Not much imagination is required to presume that the Greek philosopher Plato (ca. 428–347 BCE) borrowed the knowledge of the five regular polyhedra from Pythagoras. Plato was the first to document them in writing. Pythagoras (ca. 570–510 BCE; see bibliography, Edouard Schuré, p. 179), one of the most important founders of Greek geometry, went to Egypt as a young man to seek an education and spent about twenty-two years there.

Platonic Solids

Platonic solids (see figure above, center) are also called "Pythagorean cosmic bodies" (Weltenkörper). As the example from Timaeus—one of Plato's late dialogues, composed after 360 BCE—shows, Platonic solids were thought to be much more than just geometric shapes: from a philosophical and spiritual point of view, they represented the basic elements for considering cosmic and divine qualities. In his dialogue Timaeus, for example, Plato makes Socrates express that God must have created the four elements and the cosmos from the five perfect Platonic solids. In his description of the composition of the world, Plato developed a complex construction of regular polyhedra from two kinds of "most beautiful" triangles. The four elements—the solid, the liquid, the volatile, and the fiery—are then related to four of the five regular polyhedra (cube, icosahedron, octahedron, and tetrahedron). Apart from that, Plato refers to a fifth, all-encompassing element. But the analogic description of the dodecahedron is absent. In another philosophical dialogue (Phaedo), Plato has Socrates express that in a "transfigured meditation" the earth has an appearance alike to "leather balls with twelve parts."

Roman dodecadra and icosahedron

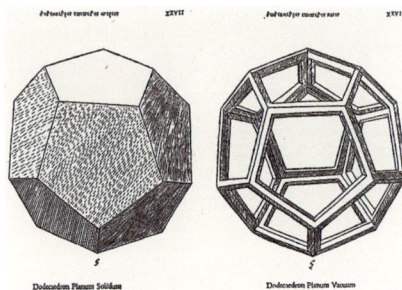

Dodecahedron, illustrated by Leonardo da Vinci

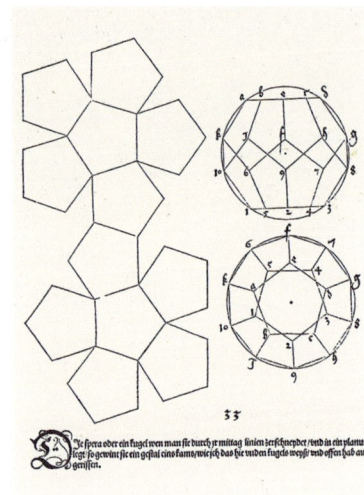

Flat pattern of a dodecahedron
by Albrecht Dürer

Archimedean Solids

The Greek mathematician Pappos of Alexandria (ca. 320 CE) stated that the scholar Archimedes (ca. 287–212 BCE) had provided a complete description of the series of the fifteen semiregular ("Archimedean") solids. These can be constructed from two or three different regular polygons. The Greek mathematician Euclid (360–280 BCE) in his "Elements"—a comprehensive systematic treatise on arithmetic and geometry—treats all Archimedean solids (see glossary, p. 176). The geometric achievements of the Greeks survived in the Arabic world during the long centuries when Europe no longer knew what to do with geometry or, rather, not yet, and when occasionally the church opposed its study. Many an ancient work had to be retranslated later from Arabic.

Dodecahedra and Icosahedra around the Year

Around the turn of the era, only two material forms were known of the polyhedra described here. The first is a dodecahedron, with round openings providing a transparent view and with nodular, bud-like attachments at the vertices. Another form from that period is an icosahedron, with slightly concave, closed sides and vertices similarly embellished with nodes. Whether these shapes, made from bronze, had any meaningful use is still the subject of much heated debate (see figure on the left).

Illustrations by Leonardo da Vinci

The Renaissance eventually also breathed new life into the classical concepts of geometry. The mathematician Luca Pacioli (ca. 1445–1517) and Leonardo da Vinci (1452–1519), as author and illustrator of Divina proportione in congenial cooperation, published a first textbook of mathematics in the Italian language in 1509. It contains all Platonic solids, as well as some of the Archimedean solids and other star polyhedra. Each spatial figure is illustrated twice: first as a closed solid, and on the following page as an open stick figure (see figure, center).

Flat Patterns by Albrecht Dürer

Only a few years later, in 1525, a first textbook in German appeared with Underweysung der messung mit dem zirckel und richtscheyt in Linen ebenen und gantzen corporen ("Instruction for measuring lines, planes, and entire solids with the compass and ruler"). Albrecht Dürer (1471–1528), familiar from his journeys with the knowledge of his Italian colleagues, included the Platonic and some Archimedean solids in his book. To facilitate reproduction, all models are illustrated in flat patterns (see figure on the right and glossary, p. 177). Not long afterward, goldsmith and copper engraver Wenzel Jamnitzer (ca. 1507–1585), from Nuremberg, illustrated a number of polyhedra models in his work Perspectiva corporum regularium (1585).

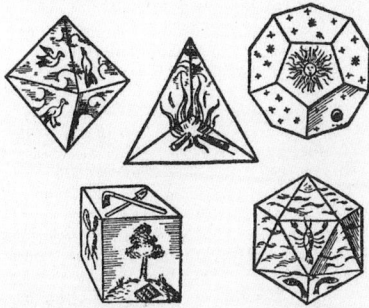

Keplers Darstellung der fünf regelmäßigen Körper
(später im zweiten Buch der «Weltharmonik»)

In his illustration of the five regular polyhedra, Kepler takes up Plato's association with the cosmic elements.

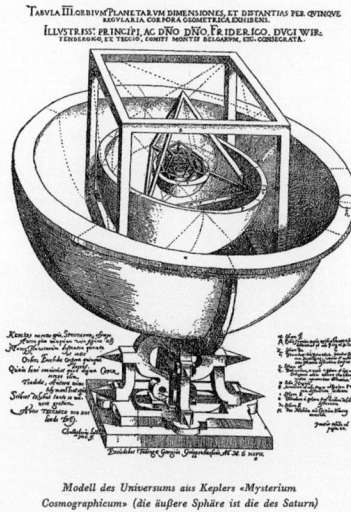

Modell des Universums aus Keplers «Mysterium
Cosmographicum» (die äußere Sphäre ist die des Saturn)

Platonic solids as space keepers between the planets
in Kepler's *Mysterium cosmographicum*

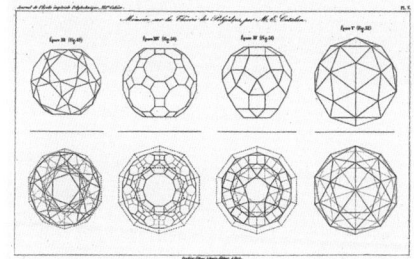

Catalan solids were exhaustively described around 2,000 years after the Archimedean solids.

Kepler's Platonic Solids

Astronomer Johannes Kepler (1571–1630) stood at the threshold of the modern era. Throughout his life, he maintained an enthusiasm for the classical ideas of a harmonic order. His illustration (see figure on the left) shows the regular polyhedra as described by Plato, with the attributes of the cosmic elements. Apart from that, he discovered some star forms of the Platonic solids. The star tetrahedron is also called a "Kepler star" (or "stellated octahedron").

Mysterium cosmographicum (1597)

As an astronomer with a strong sense for scientific measuring, and as a human being with a longing for a philosophical and religious world order, Kepler tried to gauge the reasons for the distances between the planetary orbits around the Sun through geometry. In the midst of a mathematics class he had to teach in Graz, he suddenly thought that the Platonic solids could have served God as separators between which the planets follow their nearly circular paths on individual spheres (see figure, center).

According to Kepler, the Mercury sphere is surrounded by an octahedron, followed by the sphere of Venus (icosahedron), that of the Earth (dodecahedron), the Mars sphere (tetrahedron), the Jupiter sphere (cube), and finally the sphere of Saturn. Exact measurements that were made later deviate from Kepler's model, but at least he achieved a rather remarkable approximation of reality. At the time, the planets beyond Saturn were not yet known. Incidentally, Kepler's model does not explain the distance between the Sun and Mercury.

The Modern Era

The engagement with geometry was dominated by concepts deriving from primary, sensory, visual impressions until, with the tendency toward abstraction that started during the mid-nineteenth century, an entirely new worldview emerged. The realizations gained earlier from direct observation were now used as a basis to draw conceptual, logical conclusions on which our modern scientific worldview rests. Sensory observation appeared to serve merely as an occasion for an examination of a particular topic, and it can be trusted only when it remains without contradiction from a conceptual, logical perspective.

However, no atom derives from sensory observation, since atoms cannot be seen or felt. Nevertheless, spatial structures can be scientifically recorded and classified. The exact examination of chemical processes leads to the discovery of examples of lawfulness that are best explained by the assumption of the existence of molecules and a concept of them as existing in spatial structures. The periodic system of the elements was discovered by Russian chemical scientist Dmitri Ivanovich Mendeleev (1834–1907) and independently shortly afterward by his German colleague in the same field, Lothar Meyer (1830–1895), who eventually published in 1864. The classification of crystalline forms in the mineral kingdom is a good example of such spatial structural models of thought.

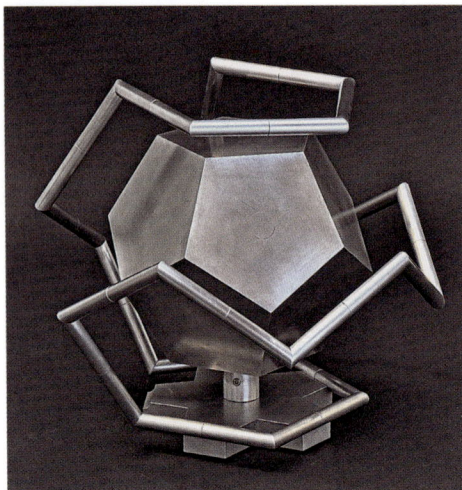

Dodecahedron with Hamilton cycle

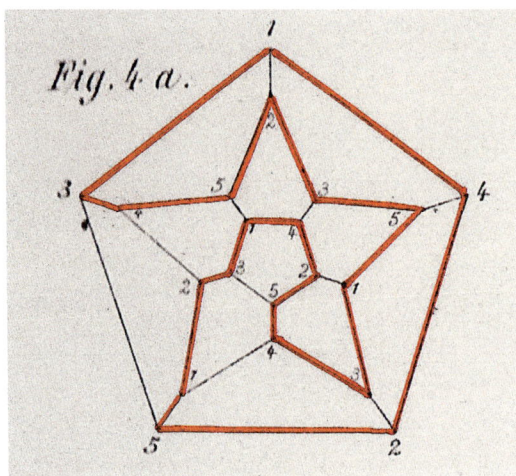

Schlegel diagram with Hamilton cycle

This new possibility of abstract thinking also became visible in the illustration of polyhedra from the mid-nineteenth century. Examples of this are the molecular spatial structures of the American winner of the Nobel Prize for Chemistry, Linus Pauling (1901–1994), as illustrated in his *The Architecture of Molecules*.

Around 1850, Swiss mathematician Ludwig Schläfli (1814–1895) was working on a paper in which he examined spatial structural laws. In this work he also abstracted spatial polyhedra forms and depicted them in formulae. The dodecahedron, with its hexagonal surfaces and its vertices where three edges meet, is designated by Schläfli as a three-dimensional form with face brackets as {5, 3}. During his own lifetime, he received little appreciation from mathematicians for his work. His work *Theorie der vielfachen Kontinuität* ("Theory of multiple continuity") was published posthumously. A historical review shows that this publication lays important theoretical foundations for spatial structures. These are still valid today.

During the same period, Irish mathematician and physicist William Rowan Hamilton (1805–1865) was occupied with the question of how a closed path along the edges of a polyhedron could touch each vertex of the polyhedron only once. There are several possible solutions for many polyhedra shapes. Such paths have since been called "Hamilton cycles" or "Hamilton circuits." In the figure above, *at right*, a Hamilton cycle has been drawn in orange onto a Schlegel diagram. The sculpture by Friedhelm Kürpig, formerly professor of geometry at the University of Fine Arts (HFBK) in Hamburg, demonstrates this connection very clearly in a dodecahedral form (see fig., *left*). The two-dimensional illustration of polyhedra enabling the depiction of all vertices, edges, and side surfaces was named after German mathematician Victor Schlegel (1843–1905). The advantage of this form of depiction is that no surface of the polyhedron covers another. The surfaces are, however, grossly distorted. The Schlegel diagram (*above*) provides a view from the surface of a hexagon into the inside of a dodecahedron. Each one of the eleven hexagonal surfaces on the inside of the dodecahedron is visible. And together with the "window" that provides the view, there is a total of twelve surfaces.

Small replica of the Atomium at Minimundus, Klagenfurt, Austria

Geometrical art, Basel, Switzerland (near the SBB railway station)

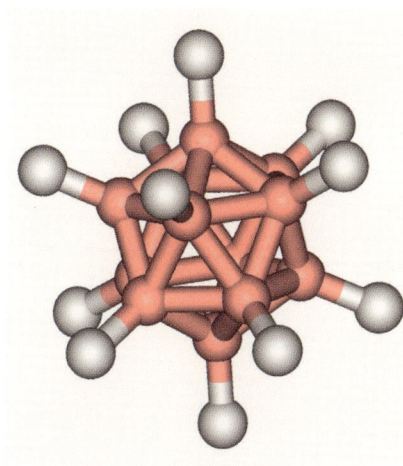

Icosahedron form: dodecaborate anion

Catalan (or Dual-Archimedean) Solids

The discoveries of the modern era and the possibility of abstract thinking prepared the context in which, 2,000 years after Archimedes, Eugène Charles Catalan (1814–1894) successfully discovered a second series of semiregular solids. Each Archimedean solid has a dual form. Catalan was the first to describe all thirteen dual forms in their entirety, and he eventually also mathematized the Platonic and Archimedean solids. His illustration is full of mathematical formulas. But at the end of the journal in which his work of about seventy pages appeared in 1865, there are also plates of those semiregular polyhedra that are dual (i.e., geometrical "twin forms"; see p. 13, figure on the right, and glossary, "dual partner," p. 177) in relationship with the Archimedean solids. Through this dual relationship they are most closely related to the Archimedean forms.

It is barely conceivable to our intellectual abilities today that these Catalan solids could not be fully described until about 2,000 years after the Archimedean solids. Our modern understanding of space seems much more acute with regard to theoretical approaches: through mathematical abstraction, we can access spatial forms today that cannot be formed by a simple joining of regular polygons.

Everyday Geometry of Polyhedra

Regular and semiregular spatial forms are part of our every-day life in a variety of areas. The spatial structural characteristics of polyhedra are used, for example, to illustrate chemical and crystallographic structures. The Atomium in Brussels functions both as a building and as an oversized model of an enlarged cubic cell of an iron crystal structure consisting of nine atoms. The entire field of modern chemistry is inconceivable without spatial structural models. In its classification, crystallography, too, relies on polyhedra forms. Fluoride is easily split into an octahedron form (see p. 16, figure on the right). And in biology, pathogens are known that have the form of polyhedra.

For the construction of roofs, climbing frames (see p. 16, figure, *center*), and radar domes (see p. 17, figure, *right*), static load-bearing capacity and an efficient modular building strategy based on planes and poles of uniform size are skillfully combined. The geodesic-dome buildings technically realized by Richard Buckminster Fuller (1895–1983) have become famous in this respect. They are an advance on the architectural solutions that were previously tried out in the building process of the Zeiss planetarium in Jena, Germany, which was opened in 1926. Today, entire stadium roofs are constructed using nodular frame structures. By using such structures, it is possible to achieve a markedly improved relation between load capacity and dead load as compared with a solid execution using lightweight construction materials.

Truncated cube clock in Klagenfurt (A), Austria market square

Icosahedron climbing frame

Fluorite octahedron

Helmut Emde (from 1972 to 1991, professor at TU Darmstadt in architecture for "mathematics for architects and geometrical information processing") has conducted research into regular and semiregular polyhedra forms to create a geometrical basis for these construction techniques (see bibliography, p. 178). The immediate aesthetic effect of the polyhedron is just as important in climbing frames as it is in lampshades or public clocks. An observant eye will soon find examples in its everyday environment. In Austria, for example, there are often clocks in the form of truncated cubes in larger public squares (above left). Rounded out into a sphere, regular and semiregular spatial forms become the basis for the design of sports balls, such as in the cutting of the individual pieces or the graphic design of printed-on patterns. Whatever the makers of sports products present to us as their newest ball for the game, they cannot avoid using the forms of platonic solids as the structural basis for their design (see p. 11, central figure).

The Eastern History of Paper . . .

. . . in its general outline is soon told: in 105 CE, Cai Lun (ca. 50–118 CE), official at the Chinese imperial court, describes the process of papermaking—the first historically verified docu-

ment. Archeological findings support the fact that paper must have existed as early as the second century BCE. But only around the time of the turn of the first millennium did Arabs bring the knowledge of making paper to Europe. The first paper mills in central Europe were then built shortly before the start of the fifteenth century.

The simplest forms of folding paper must go back to the invention of the material itself. Simple origami forms probably originated in ceremonial paper folding as practiced by Japanese monks. In the 1960s, origami enjoyed an increasing popularity that spread rapidly, and numerous new variations were created. The models presented in *Folding Polyhedra* can be seen as a special branch of origami. Geometrical forms from joined modules are also called "modular" or "mathematical origami," a rather recent direction of this Japanese folding technique.

No other material is suited so well for folding polyhedra as paper. The fiber structure of paper is even but chaotic: the individual paper fibers lie arbitrarily distributed across the plane without any regularity, as opposed to the regularity of spatial polyhedra structures. Even in this respect, the relationship between the two brings together opposites.

Illuminated octahedron structure in Wetter/Ruhr (D), Germany

Radar dome, pentakis dodecahedron, Berlin-Tempelhof, Germany

Paper and origami	Time frame	Geometry of polyhedra
	3000–2000 BCE (Neolithic)	Carved Stone Balls
	1550–1070 BCE	Ancient Egypt (New Kingdom), flower of life
	ca. 570–510 BCE	Pythagoras and his "cosmic bodies" (Weltenkörper)
	428–347 BCE	Platonic solids (after Plato): comprising 5 regular polyhedra
	287–212 BCE	Archimedean solids (after Archimedes): comprising 13 semiregular polyhedra
earliest datable references to paper	2nd century BCE	
	around the turn of the era	Roman forms of icosahedra and dodecahedra
Cai Lun (China) describes papermaking	105 CE	
Paper reaches southern Europe	1000 CE	
First German paper mill in Nuremberg	1390	
	Renaissance period: 1509	Divina proportione by Luca Pacioli, illustrated by Leonardo da Vinci
	1525	Albrecht Dürer's Underweysung published
	1568	Wenzel Jamnitzer: Perspectiva corporum regularium
	1597	Mysterium cosmographicum by Johannes Kepler
	1865	Catalan solids: Eugène Charles Catalan writes a treatise on dual Archimedean solids
origami boom	1960s	

Regular and Semiregular Polyhedra

Platonic Solids—Regular Polyhedra

Since it was Plato who first mentioned these in writing, regular polyhedra are also called "Platonic solids" (see p. 11, central figure). After the sphere, regular polyhedra are among the simplest and most-symmetric spatial forms. If we approach them with our imagination, it is easiest to picture them as a combination of a certain number and type of faces (e.g., triangles)—the tetrahedron from four, the octahedron from eight, and the icosahedron from twenty. To these can be added the cube, consisting of six squares, and the dodecahedron from twelve pentagons. The designations for the number of faces have their roots in Greek and Latin. With the same justification, we could describe these spatial forms on the basis of their vertices, or on the basis of the number of edges and the surfaces that meet at the edges. This would appeal to chemical scientists, among others, who like to think in spatial, modular structures. And it would also fully correspond to the approach to folding and constructing followed in this book. The number of edges would lead to just as good a basis for the designation of the individual spatial forms. Perhaps one should generally call them "figures," which privileges neither vertices nor edges nor faces, and which in its visual connotations introduces a dynamic aspect into the static world of shapes. Nevertheless, we follow the general use

of terminology and continue to use the term "polyhedra." All Platonic solids are pictured on the following spread (see p. 20) in the left column, one beneath the other with the full name of the model. Additionally, at the bottom end of the page you will find the geometrical profile of each folded polyhedron form, with a list of their essential characteristics (e.g., the number of their faces, vertices, and edges).

Cropping of Platonic Solids:
Archimedean Solids—Semiregular Polyhedra

Apart from the five regular platonic solids, there are—as already described—thirteen Archimedean, now semiregular, solids that can be imagined as consisting of two or three different areas joined together. These shapes can also be created by cropping Platonic solids at their vertices, sometimes also at their edges. This results in the truncated forms of the tetrahedron, octahedron, icosahedron, cube, and dodecahedron. This can be extended into cubo-octahedra, icosidodecahedra, and their truncated forms. To these can be added two strangely twisted solids that can be formed in two chiral shapes: the snub cube (cubus simus) and the snub dodecahedron (dodecahedron simum). The term "chiral" means that each form has two mirror-inverted variants (see glossary, p. 177).

All Archimedean solids are pictured on the following spread (see p. 20) on the right-hand side, next to the Platonic solids.

Raising of Platonic Solids:
Catalan Solids—Semiregular Polyhedra

Each Archimedean solid has a dual partner (see glossary, p. 177). This is a kind of inverted twin brother. Catalan solids can be deduced mathematically and geometrically from Archimedean solids. On page 21, they have therefore been pictured as mirror images of the Archimedean solids. It is also possible to find them by raising the center points of the faces of a Platonic solid outward (like a tent). The results are the triakis tetrahedron (from the tetrahedron), triakis hexahedron (from the octahedron), pentakis dodecahedron (from the icosahedron), and triakis octahedron (from the cube), as well as the triakis icosahedron (from the octahedron), the rhombic dodecahedron, and rhombic tricontahedron and their deltoid forms (deltoidal icositetrahedron, deltoidal hexecontahedron). With the Catalan solids we also find chiral forms (pentagonal icositetrahedron, pentagonal hexecontahedron). All Catalan solids consist of only one face shape.

Summary

There are five different regular polyhedra (Platonic solids). In addition, there are thirteen semiregular (Archimedean) and a further thirteen semiregular (Catalan) solids. In total there are thirty-one regular and semiregular spatial forms. If we add the chiral variants—when building models, we have shown both variations below—there are thirty-five forms/models in total. Icosahedra and dodecahedra can also be realized as chiral forms. This results not from necessity, but rather from a playful approach based on the similarity between these and other chiral forms. In the overview of all forms of polyhedra by Helmut Emde (see p. 177), this is printed in the margins above the dodecahedron and the icosahedron. In contrast to accepted spelling, the names of the semiregular polyhedra occasionally have been written with a hyphen to highlight geometrical relationships and to facilitate the differentiation of the individual forms from the regular polyhedra. The numbers next to the individual forms (see pp. 20–21) refer to the sequence in which the models are introduced in this book. This sequence does not primarily follow the geometrical similarities but starts with simple models and then leads step by step to higher levels of difficulty. Models with a similar structure have been grouped together. The most-important geometrical basic terms for this book have been explained separately on p. 176 and following.

Platonic Solids

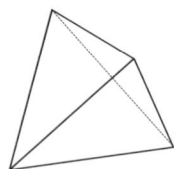 E17

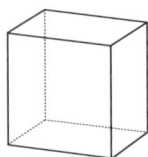 E16

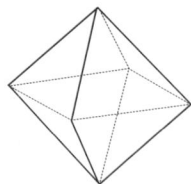 A01

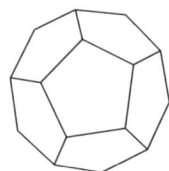 B04

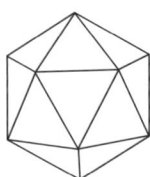 B08

Archimedean Solids

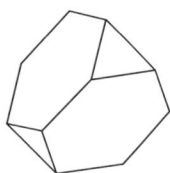 B05

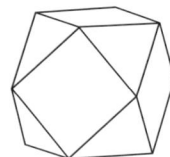 A02

 F20

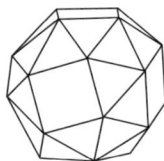 F21

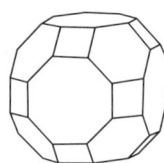 F19

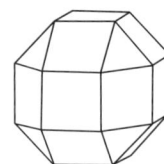 A03

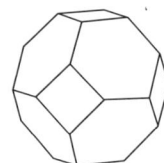 B06

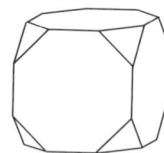 F18

 F22

 G25

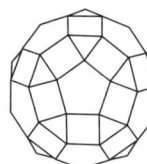 G24

 B07

 G23

E17 tetrahedron
f 4, v 4, e 6

E16 cube
f 6, v 8, e 12

A01 octahedron
f 8, v 6, e 12

B04 dodecahedron
f 12, v 20, e 30

B08 icosahedron
f 20, v 12, e 30

B05 truncated tetrahedron
f 8, v 12, e 18

A02 cubo-octahedron
f 14, v 12, e 24

F20 icosidodecahedron
f 32, v 30, e 60

F21 cubus simus (snub cube)
f 38, v 24, e 60

F19 truncated cubo-octahedron
f 26, v 48, e 72

A03 rhombic cubo-octahedron
f 14, v 24, e 48

B06 truncated octahedron
f 14, v 24, e 36

F18 truncated cube
f 14, v 24, e 36

F22 dodecahedron simum
f 92, v 60, e 150

G25 truncated icosidodecahedron
f 62, e 120, e 180

G24 rhombic icosidodecahedron
f 62, e 60, e 120

B07 truncated icosahedron
f 32, v 60, e 90

G23 truncated dodecahedron
f 32, v 60, e 90

Catalan Solids

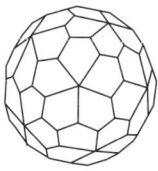 **D15**

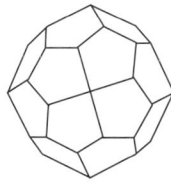 **D14**

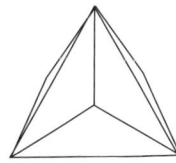 **H27**

 **H31**

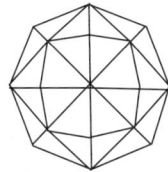 **H29**

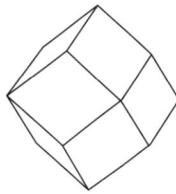 **C09**

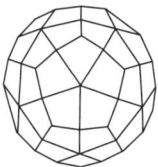 **D13**

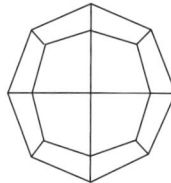 **D12**

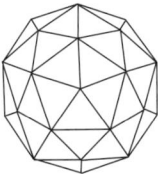 **C11**

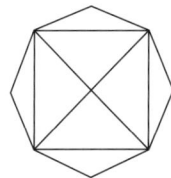 **H26**

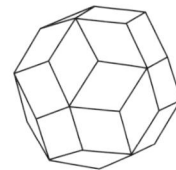 **C10**

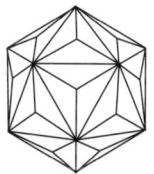 **H30**

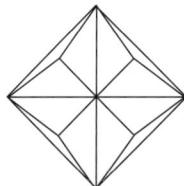 **H28**

D15 pentagon hexecontahedron
f 60, v 92, e 150

H31 pentakis dodecahedron 4/6/10
f 120, e 62, e 180

D31 deltoidal hexecontahedron
f 60, e 62, k 120

C11 triakis icosahedron
f 60, v 32, e 90

H30 pentakis dodecahedron
f 60, v 32, e 90

D14 pentagon icosatetrahedron
f 24, v 38, e 60

H29 triakis hexahedron 4/6/8
f 48, v 26, e 72

D12 deltoidal icosatetrahedron
f 24, v 14, e 48

H26 triakis octahedron
f 24, v 14, e 36

H28 triakis hexahedron 3/8
f 24, v 14, e 36

H27 triakis tetrahedron
f 12, v 8, e 18

C09 rhombic dodecahedron
f 12, v 14, e 24

C10 rhombic tricontahedron
f 30, v 32, e 60

Key:
The numbers following the letters A, B, C, D, E, F, G, and H refer to the sequence in which the individual models are explained in the following chapters.

The letter before the number refers to a particular group of models that share similar technical steps in their execution.

These groups correspond to chapters A–H.

Underneath each model designation, the geometrical profile of each form is given. Here I have provided the number of surfaces, vertices, and edges of each model:

f = number of faces/surfaces
v = number of vertices
e = number of edges

Practical Considerations

Paper: Quality and Color

For the kind of modular folded polyhedra presented in this book, a paper that is well sized, smooth, and dimensionally stable, with a grammage of 120 gsm, such as is often used for book covers or as flyleaf, works best. It is not always possible to obtain such papers in small quantities without incurring an extra charge, but it may be possible to find these, or similar kinds of paper, in a local printer or bookbinding workshop.

A strong, well-sized, dimensionally stable multipurpose paper is a simple alternative to premium color papers. This, too, can be used for the construction of models in the sizes given here. It can be identified by its smooth surface, and it feels notably stronger than, for example, photocopy paper. This or similar paper is also used for good-quality maps, or for envelopes made from recycled maps. Here, too, I recommend a grammage of 120 gsm. Simple printer paper, but equally traditional origami paper, is not dimensionally stable enough for the construction of these types of models.

Origami paper is normally thinner and softer and is not very well suited to the construction of models as instructed in this book. Considering the design language expressed in the polyhedra models, a paper with a neutral color is preferred to one that has a printed pattern. White paper is ideal. Choosing white with a touch of yellow will give warmth to the models. Bright white emphasizes the form more strongly. Colored models are conceivable in many variations, especially when making free forms. Constructions that reproduce the three spatial axes in different colors or, for example, illustrate two different types of modules in one model are particularly effective. If you have the opportunity to do this, you should choose colors that are subtly coordinated.

Size of the Bases

The size of the individual bases—that is, of horse and rider (in each step-by-step instruction, these are illustrated on the right side of the text) in each model—has been chosen in such a way that they can be combined with each other. The C models as well as the folded polyhedra in chapters J, K, L, and M can be realized using these measurements. In the first models, which consist of only one type of base (e.g., only triangles or squares), you can also choose the size independent of the template (models A and B as well as F, G, and I). An exception are the E models: here the triangular horse-and-rider bases have different sizes. If different bases are needed in a model (e.g., squares in combination with triangles), the ratio between the two must be taken into account: the transition between a module and its neighboring module has to be of the same size. In the table "Size of the Bases for the Individual Forms of Polygons" (see p. 32), you can find all polyhedra models at a glance.

Templates

The 1:1 templates of all forms of polygons used in *Folding Polyhedra* (triangles, squares, pentagons, hexagons, octagons, and
decagons) can be downloaded at **www.schifferbooks.com/po-
lyhedra**. The best way to cut out these templates is explained.
If required, the templates can easily be enlarged or reduced to
any size by using the zoom function on your photocopier or PC.
If the individual bases are to remain mutually combinable, all
templates have to be enlarged or reduced by the same factor.

Special Measurements for Complex Models

In models C, D, E, and H, not all horse-and-rider bases are of
the same size. For the required size of the bases, please refer to
the sizes given in the table "Size of the Bases for the Individual
Forms of Polygons" (see p. 32). In this table, the length of the
edges of the base of each model has been listed summarily. The
templates for the centaur models in chapters F and G can be
downloaded and treated in the same way.

Uniform Model Size

The goal was to realize all models in a uniform size of 1.5
liters per volume. To achieve this, the measurements of each
individual base (polygon form) had to be determined through
calculation, construction, or—where this did not lead to the
desired result—trial and error. In the table "Size of the Base for
the Individual Forms of Polygons" (see p. 32), all edge lengths
of the base modules have been given in inches (") and, if necessary, slightly rounded off. In contrast to this, models J, K, L, and
M all have been given as 1:1 templates.

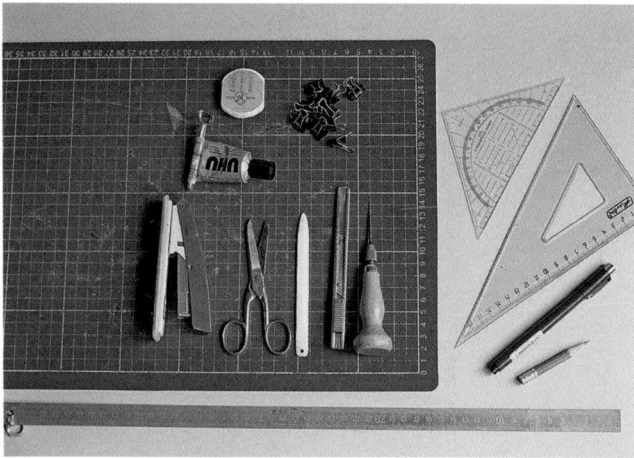

Fig. 1

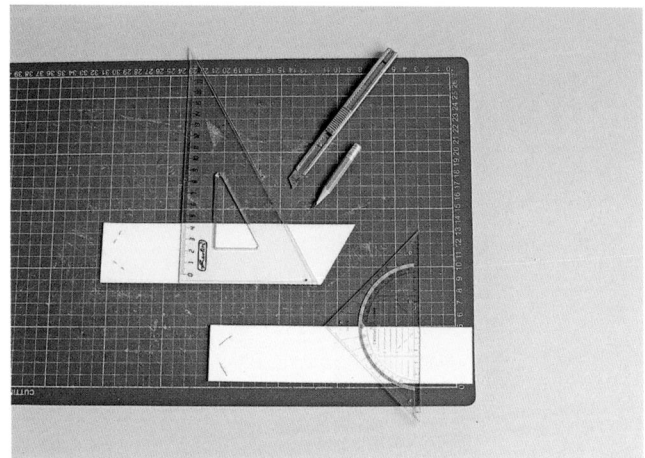

Fig. 2

Transferring the Templates and Cutting the Base

When you're cutting, it makes sense to work on several layers of paper at once. With a little bit of practice, it is possible to cut about ten sheets of paper in one go. To do this, several sheets of paper are stapled together. Staple the template on top of the other sheets. Using an awl, transfer all corner points of the template onto the sheet beneath. The template page is then folded back, so that the points (holes) can be joined by a line (see facing page, fig. 4). If you want to preserve your plastic ruler or triangular ruler, you can use a steel ruler for cutting out the polygon shapes—from one point to the next—along the pencil lines. A cutting mat saves your worktable from damage by the awl and cutter (see fig. 1). When you're cutting, it will be impossible to cut through the entire stack of paper at once: the knife has to be led along the ruler several times. Take care that the cut you are making is absolutely vertical. The blade usually has a slight bezel (or, from the sharpening surface, a slight deviation), so you have to hold the knife slightly at an angle to achieve a vertical cut.

Cutting Squares

For squares, it is easier to cut a longer strip first (see fig. 2). This is easier with a paper-cutting machine than by hand. Each strip must be as wide as the length of the sides required for the squares. The individual squares are then cut off from these strips. A right-angle ruler is best used for this.

Cutting Triangles

Strips are also needed for the cutting of triangles. The required width in this case is the height of the desired triangle—not the side length! The side lengths are given in the table "Size of the Bases for the Individual Forms of Polygons" on p. 32. You can work out the height through simple measurement. You can then cut the 60° diagonal cuts with the help of an acute-angled set square (see facing page, fig. 3). Be careful not to cut into the set square with your cutter.

Cutting and Folding the Supports

For models H and G25, supports are needed that are pushed between horse and rider to join and stabilize two neighboring modules from the inside. Overhead projector foil is well suited to this task. The 1:1 template for the supports can also be downloaded from the internet at **www.schifferbooks.com/ polyhedra**. Their measurements can be transferred in the same way as explained for the bases. You can easily make the grooves with an awl. Along these grooves, the supports are folded "mountain-wise," as the red signature in the images shows. With a little bit of practice you can also cut and groove the supports freehand.

Fig. 3

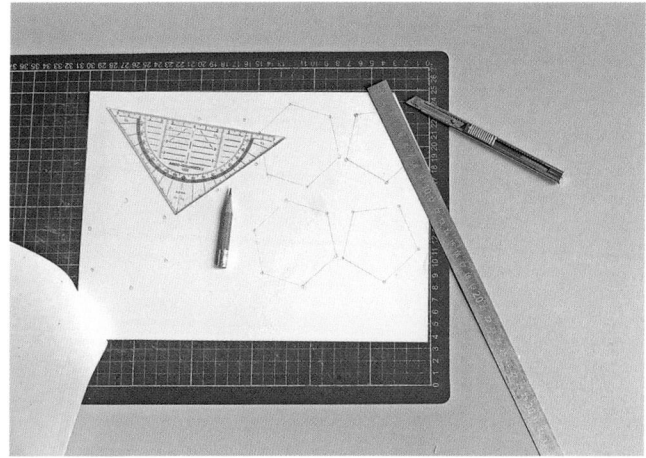

Fig. 4

Simple Tools That Can Be Helpful: Glue and Paperclips

The modules for the simple models A, B, and C, as well as occasionally other models and the extensions in chapters J, K, L, and M, can, when executed exactly, be joined into complete models without encountering any problems. For different reasons, it may be necessary to glue down the horse corners that have been folded over, or to lock the joints of two neighboring modules by using a little bit of glue. It is best to use a glue with synthetic solvents. Water-based glue is not recommended for this purpose. When the water in the solution evaporates, the paper can warp and the model can lose its level surfaces.

The smaller the area that is to be glued together, the more urgently the use of metal clips is recommended. With a metal clip, the glue can dry and harden under pressure. This has not always been explicitly included in the instructions. In shapes where glue and clips are indispensable, this becomes clear in the pictures accompanying the respective instructions.

Folding and Cuts

The necessary folds can be done by hand, or by using a folding bone. For most of the models, a small folding bone will be sufficient. Take your time with your first horse-and-rider base papers and modules, and be very careful to do each fold accurately and exactly to measure. Whether or not you will be able to successfully join the model depends entirely on ac-

curate folds and exact measurements. This applies particularly to the centaur models (chapters F and G). These have additional cuts that have to be executed as exactly as the folds in any modules. The joining of two centaur bases into a module initially requires some patience, and you have to be willing to experiment to find out how exactly the two belong together. Once the correct solution is found, it can simply be repeated for any further module required.

You Will Need:

- multipurpose paper, well sized, 115–120 gsm (see p. 22)
- overhead projector foil for supports
- pricking awl (as used in bookbinding) with a thin tip
- stapler
- paper-cutting machine (if available), cutter, and knife
- cutting mat
- steel ruler with a millimeter scale
- sharp pencil or mechanical pencil, 0.5 mm
- template
- set square and acute-angled set square
- fold-back clips and paperclips
- solvent-based glue

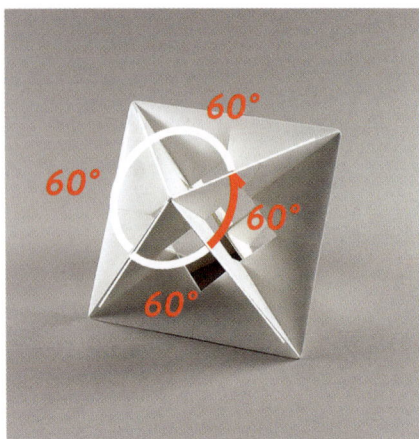

Angles in the completed model

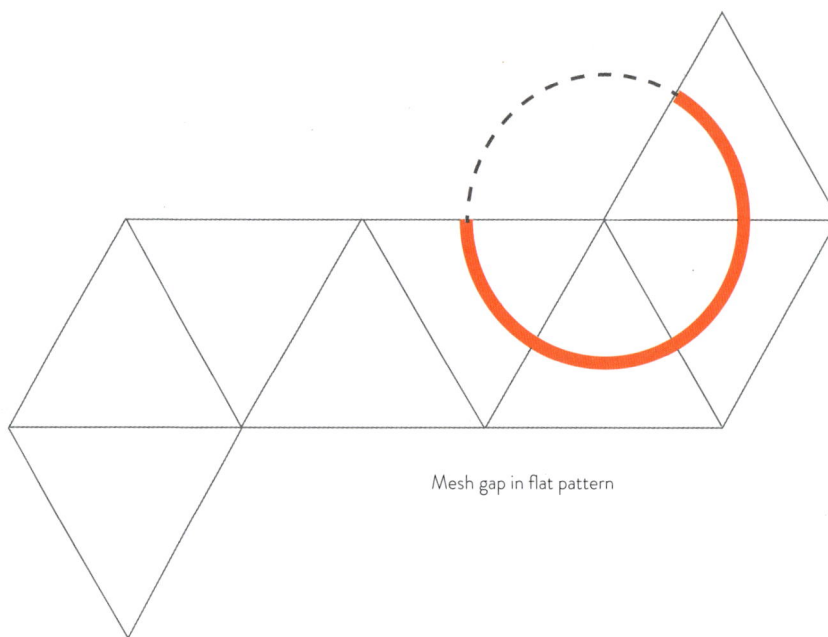

Mesh gap in flat pattern

From Polygon Center to Polyhedron Vertex

How is it possible to create the vertex of a polyhedron from a level piece of paper? This question seems trivial, but when we look closer, we see that this process is most remarkable. Initially the paper has to be folded from its position in the horizontal into space. In this process, it becomes obvious that there will be surplus material. The following consideration can illustrate this: if you draw a circle around the center of the base area with a compass, you go through an entire 360°.

Four triangles meet at each spatial vertex of an octahedron. If we follow the above compass journey, we go through 60° for each triangle (see fig., *left*, and facing page, *left*), in total 120°. Of the total of 360°, 120° are not needed and are superfluous.

Mesh Gap in Flat Pattern

If we project an octahedron into a flat pattern, the superfluous 120° would be called a mesh gap (see fig., *right*, and glossary, p. 177). The superfluous parts of paper have to be cut off, unless you are dealing with a model joined by gluing together the individual faces when any gluing flaps have to be kept intact.

In the folded octahedron (model A01; see p. 38), we initially need four mountain folds, which will form the edges of the po-

lyhedron. Between them, we have an excess on each of the four sides of 30°. For geometrical reasons, the calculated total excess of 120° is spread across the four sides. With the help of valley folds, the excess angles`/`excess material can be folded inward. In the step-by-step instructions, the horse-and-rider illustrations on the right of the text instructions show this process for each model. As illustrated on the facing page (see fig., *right*), the number "3" on the left shows which kind of polyhedron faces join in a vertex. In this example it is triangles. On the right-hand side of the illustration, you can read which angles are required between two neighboring edges in the polyhedron model: 60°. By adding the number—as described above—you will get the respective excess.

Each following model is introduced and explained in the way described. In the illustration on the left-hand side of the step-by-step instructions, we meet with regular or semiregular polyhedron surfaces, depending on the form of the respective polyhedron. The angles in the illustration on the right-hand side are in some cases given as approximate values. For the current purpose this is entirely sufficient. Those who would like to know more will find every detail in Robert Williams's publication (see bibliography, p. 180).

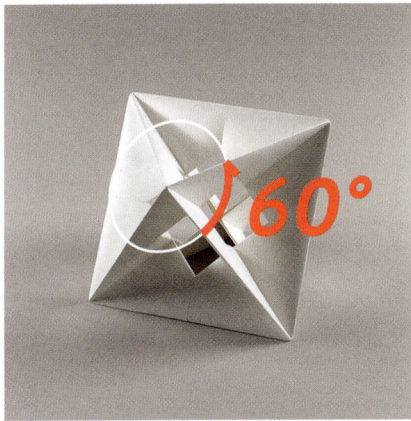

Module vertex as a meeting point of four triangles (à 60°) highlighted in the picture

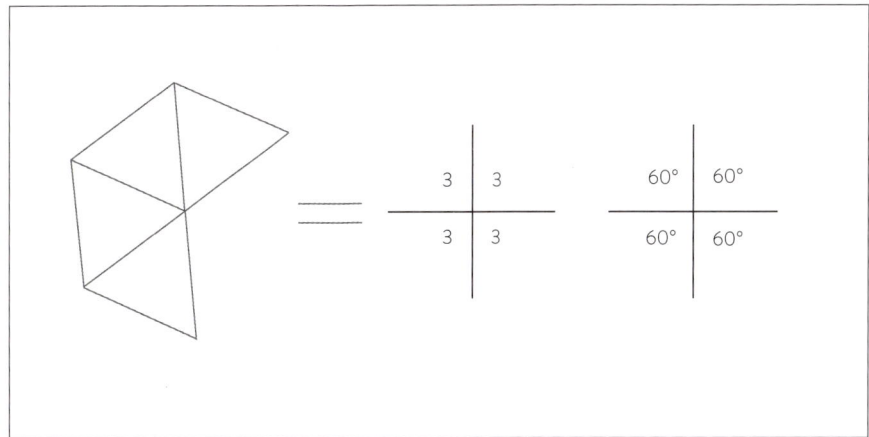

Module vertex as a meeting point of four triangles (à 60°) schematically illustrated

Horse and Rider Form a Module

A connection between two modules is possible only because parts of one module can be pushed into the slight gap between horse and rider of the neighboring module. Each module consists of two sheets; its outside is folded around the inside. The upper sheet "rides" the one below and is thus called the "rider," and the one below is the "horse." The rider is visible in the finished model (except for the tips, which are folded around the horse), and together with the remaining riders reveals the polyhedron form. The horse is no longer visible in the finished model. Within the construction it forms the core, or scaffolding, and serves the connection between the individual modules as the protruding horse tips of one module are pushed into the thin gap between horse and rider of the neighboring module. Depending on whether we are folding a horse or a rider, the distribution of mountain and valley folds is entirely different. They have to conform to the specifications discussed with reference to the excess-angle degrees. In an octahedron, the excess-angle sum is distributed evenly across the four sides between the edges of the polyhedron. As described, this is reflected in the illustrations on the right-hand side of the instruction text for each model. But some bases can function both as horse and rider. In Greek mythology, hybrid creatures of man and horse were called "centaurs." This name is used here by analogy. What this means in detail you will learn in chapter F (see p. 80 and following), which treats folded polyhedra where one piece of paper assumes both horse and rider roles.

Several modules are joined via the protruding tips of their horses into a full ring. Each "ring," of course, remains angular here. It has three, four, five, six, eight, or ten vertices and runs on straight paths between the vertices along the edges of the polyhedron. Additionally, each ring encloses a face of the polyhedron and shares exactly one vertex with the adjacent module. At the top of the base illustrations, step-by-step instructions schematically describe the number of vertices of each ring that joins a module (in an octahedron, there are four rings with three vertices each).

A second illustration shows the angle in which the edges in the rings meet in the module. In an octahedron there are four times 60°. In this sense, all modules that are needed have to be joined to the first ring. In an eight-faced octahedron, this results in a total of eight rings.

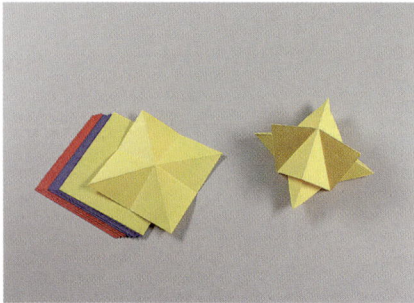

First steps: folding the bases and constructing a module from horse and rider

Joining two modules

Joining three modules into a trinomial ring joint

Note:
In the Platonic models (A01, B04, B08, E16, and E17), there is only one type of module and one type of ring joint in any one model.

Rhythmic Assembly
On the completed model, the eye can travel from one vertex to another. In regular models, the step to the next neighboring vertex is always the same. In the semiregular models, there is an inherent rhythm that is different for each model. In a rhombic dodecahedron (model C09; see p. 58), for example, a quadrinomial vertex is always followed by a trinomial vertex.
In this sense, such visual journeys from vertex to vertex, or even better with the finger, always follow a repeating rhythm that also emerges in counting neighboring or successive modules. This rhythm also facilitates orientation when assembling the model. You can also use the accompanying pictures of completed models to orient yourself.

Note:
In the models of Catalan solids, there are always two or three different modules, with only one type of ring joint in one model.

In Archimedean solids, there is only one type of module, with two or three different types of ring joint in one model.

To make it easy to get started, the first simple folded polyhedra are initially introduced in two or three different colors. This emphasizes the borders between the vertex modules. In addition, colored models are far more decorative than white models. In some circumstances, there are—depending on whether a model is realized using two or three colors—different sequences of assembly for the same type of polyhedron.

Naming the Polyhedra Models:
A Principle of Letters and Numbers
In addition to its geometrical name, each model in *Folding Polyhedra* is designated by a combination of letters and numbers. The running numbers follow the level of difficulty, from simple to complicated models. The letter preceding the number refers to closer similarities of several models that have been grouped together in one chapter. The extended polyhedra forms (see p. 120 and following) have their own numbers.

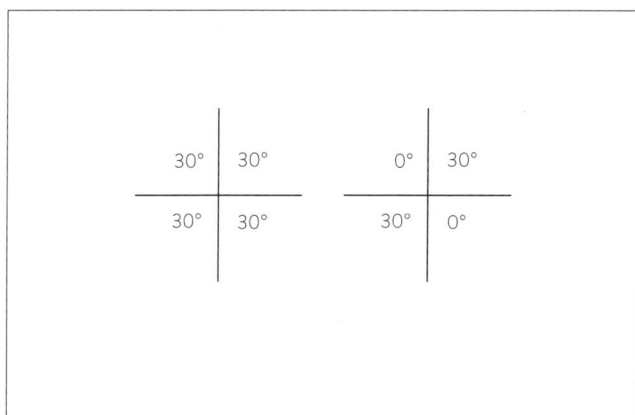

Excess angles in an
octahedron (*left*)
Excess angles in a
cubo-octahedron
(*right*)

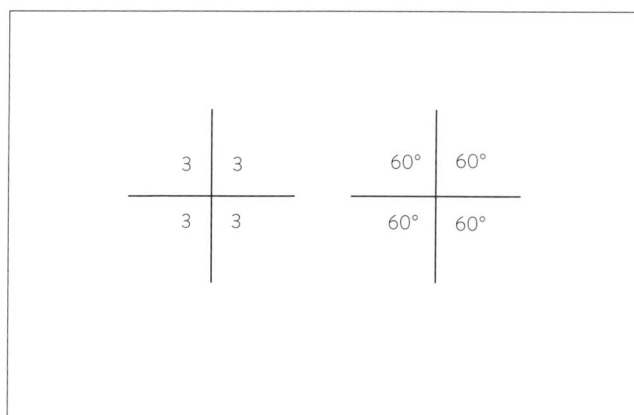

In each module in an octahedron, four triangles at 60°
degrees meet.

A	Models based on squares
B	Models based on triangles and pentagons
C	Combinations of triangles, squares, pentagons, and hexagons
D	Horse and rider of different size
E	Additional mountain folds and envelopments
F	Centaurs—combined horse-rider function
G	Complex centaurs
H	Models with supports
I	New chiral forms
J	Extended cubo-octahedra and rhombic cubo-octahedra
K	Extended dodecahedra
L	Extensions of truncated tetrahedra, octahedra, and icosahedra
M	Extensions of octahedra, icosahedra, and rhombic dodecahedra

Terminology and Key to Illustrations

I have used a uniform system of illustration throughout the book for all relevant details of the folded polyhedra models. In addition to the step-by-step instructions for each model, there is a visual illustration of the number of edges that meet in a vertex, and the form of the faces that are being defined by the edges of the polyhedron. Next to this is another illustration that shows the required angles between the edges of the polyhedron.

Below this, the horse-and-rider bases with their required mountain and valley folds, as well as—if applicable—cuts (centaurs), are shown. Here the rider is always shown on the left side, the horse on the right.

outline of the sheet of paper: black line

mountain folds (edges of the polyhedra): red line

valley folds (usually serve to fold excess angles into the polyhedron): dotted line

cuts (in centaurs): double line

Polygon Shape

The starting point is always a level surface or a sheet of paper with several corners (3, 4, 5, 6, 8, or 10). These base surfaces function either as horse or rider (see p. 30), or both (in centaurs). In the models here, only regular polygons are used (i.e., polygons with uniform angles and edge lengths).

Mountain fold ——————

Valley fold --------------------------------

Mountain fold

Valley fold

Mountain fold

Rider (top)

Horse (bottom)

Mountain fold ——————

Valley fold --------------------------------

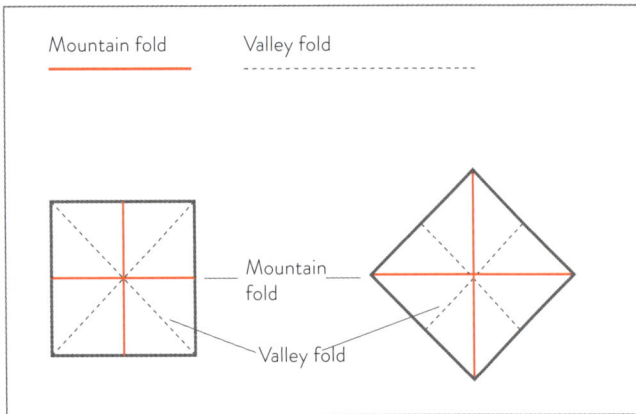

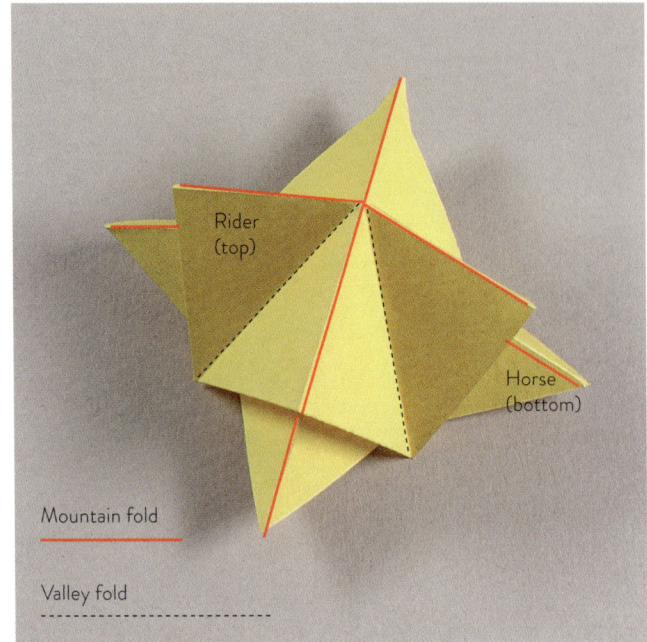

The Shape of the Polyhedron

Folded polyhedra are multifaceted spatial forms. In regular polyhedra, all angles, edge lengths, and faces of the polyhedron are the same; in semiregular polyhedra, all remaining spatial forms. The extended polyhedra form a separate group.

Pictograms

The horse symbol shows that this area is at the bottom of a module. The rider "sits" on top, and protruding tips are folded around the horse (see facing page, fig. at the top). In the centaurs this is similar. You can find detailed instructions in chapters F and G (see pp. 80 and 92).

 Horse   Rider   Centaur

horse: The polygonal base in any module/model that always lies inside, serves as a link between two modules, and is not visible from the outside of the module

rider: The polygonal base in any module/model that always lies outside, whose corners/tips are folded around the horse-sheet inwardly; the remaining base is visible in the final model.

centaur (ct): The polygon base with combined horse-rider function (see chapter F, p. 80 and following)

module: A horse-and-rider base (or two centaurs) together form one module. Depending on the number of protruding corners/tips that function as joints with the neighboring module, there are three-, four-, five-, six-, eight-, and tenfold modules.

ring joint: Several modules can be joined in a ring or put together in a ring joint. Each ring is in itself angular and runs between the vertices along the edges of the polyhedron on straight paths. Additionally, each ring circumscribes a surface of the polyhedron. The number of edges in the polyhedron is always provided on the right-hand side of the step-by-step instructions. In an octahedron there are only triangular surfaces, and thus only triangular ring joints.

rhythmical paths: It is possible to "walk" from any vertex of a polyhedron to the next with the eyes or with the fingers. In doing so, a number of paths emerge. Each model has its own rhythms. In an octahedron, for example, there is an inherent threefold rhythmical path around a surface, and a fourfold path once around the entire model.

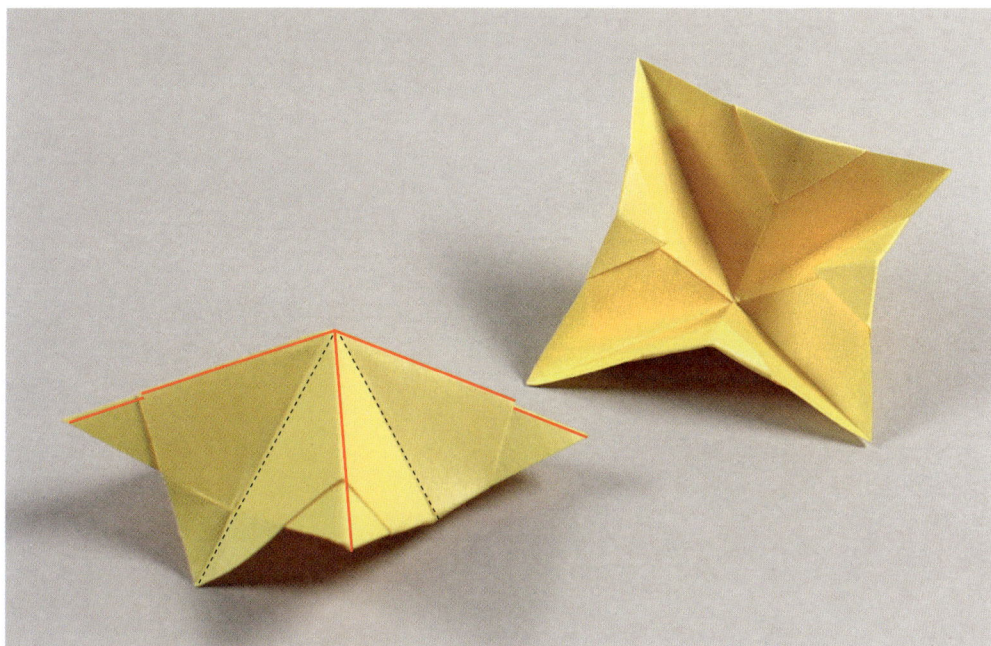

Folding the rider tips inwardly around the horse edges

All models and ring joints are put together in a rhythmically repeating sequence. Each model has its own rhythm. To understand this rhythm, it is enough to follow the edges of the models (see appropriate illustrations for each model).

model: A spatial form consisting of several modules (polyhedra form)

Forms

The bases consist of regular triangles, squares, pentagons, hexagons, octagons, or decagons. All polygonal forms used in this book are regular, even if this is not explicitly stated. None of the models have been folded from other forms. The shape of the base required for each model is illustrated—including mountain folds, valley folds, and, if required, cuts—with the step-by-step instructions of the individual folded polyhedra.

Size of the Bases

Owing to the intended uniform size of the folded polyhedra (as well as for other reasons), the bases for each polyhedron form in chapters A–I are of different sizes. In the extended models J–M, they have the same size. You can find the required length of the edges of the bases in the table "Size of the Bases for the Individual Forms of Polygons" (see p. 32):

Example 1: Octahedron (Model A01; see p. 38)
To build an octahedron, squares with an edge length of 6" are needed (horse and rider are thus of the same size). In the table, you will find the required size of the squares in row "A01," column "square."

Example 2: Deltoidal icositetrahedron
Building a deltoidal icositetrahedron (row "D12" see page 32) requires triangles. The horse triangles have an edge length of 2.6" (column "triangle," "horse"), while the rider triangles have an edge length of 2.3" (column "triangle," "rider"). The edge length of the squares for horse and rider is 2.35".

From Template to Required Size

All polygon illustrations are given in the step-by-step instructions in the same size. In the templates that you can download for each model (see p. 23), the bases of horse and rider are printed on a scale of 1:1. In the following table, all edge lengths of the starting modules for chapters A–I have been given in inches (") and, if necessary, are slightly rounded off. In contrast, the templates for models J–M were uniformly realized on a scale of 1:1.

Base Sizes for Each Polygon Shape

No.	polyhedron	Triangle		Square		Pentagon		Hexagon		Octagon		Decagon	
		Required Edge Length in Inches											
		horse	rider	horse	rider	horse	rider	horse	rider	horse	rider	horse	rider
A01	octahedron [square] 6			15									
A02	cubo-octahedron [square] 3.38			8,6									
A03	rhombic cubo-octahedron [square] 2.1			5,5									
B04	dodecahedron [triangle] 4	10,3											
B05	truncated tetrahedron [triangle] 5.6	14,2											
B06	truncated octahedron [triangle] 3.75	9,5											
B07	truncated icosahedron [triangle] 2	5,2											
B08	icosahedron [pentagon] 2.6					6,6							
C09	rhombic dodecahedron [triangle] 4.45	11,3		9,2									
C10	rhombic tricontahedron [triangle] 2.5	6,3				4,6							
C11	triakis icosahedron [pentagon] 1.19					3,0		3,2					
D12	deltoidal icosatetrahedron [triangle] 2.5	6,6	5,8	6,0									
D13	deltoidal hexecontahedron [triangle] 1.45	3,7		3,1		3,5	3,7						
D14	pentagonal icosatetrahedron (chiral) [triangle] 2.32	5,9		5,8	6,1								
D15	pentagonal hexecontahedron (chiral) [triangle] 1.3 [pentagon] 1.3	3,3				3,3	3,6						
E16	cube [triangle] 7.83	19,9	20,1										
E17	tetrahedron [triangle] 16	40,6	41,0										
F18	truncated cube [square] 1.88 ct			4,8 zt									
F19	truncated cubo-octahedron [square] 1.3 ct			3,3 zt									
F20	icosidodecahedron [pentagon] 1.37 ct					3,5 zt							
F21	cubus simus (snub cube; chiral) [hexagon] 1.3 ct							3,3 zt					
F22	dodecahedron simum (chiral) [hexagon] 0.75 ct							1,9 zt					
G23	truncated dodecahedron [pentagon] 0.75 ct					1,9 zt							
G24	rhombic icosidodecahedron [hexagon] 0.75 ct							1,9 zt					
G25	truncated icosidodecahedron [hexagon] 0.5 ct							1,2 zt					
H26	triakis octahedron [square] 1.89			4,8				5,4	5,6				
H27	triakis tetrahedron [triangle] 4.6	11,7	8,4					11,1					
H28	triakis hexahedron 3/8 [triangle] 2.48	6,3	4,6							5,8			
H29	triakis hexahedron 4/6/8 [square] 1.3			3,3				3,3		3,5			
H30	pentakis dodecahedron 3/10 [triangle] 1.3	3,3	3,3									2,8	
H31	pentakis dodecahedron 4/6/10 [square] 0.75			1,9				1,9				1,9	
I32	icosahedron (chiral) [hexagon] 2 ct							5,1 zt					
I33	dodecahedron (chiral) [triangle] 4	10,3											

Rhombic Dodecahedron

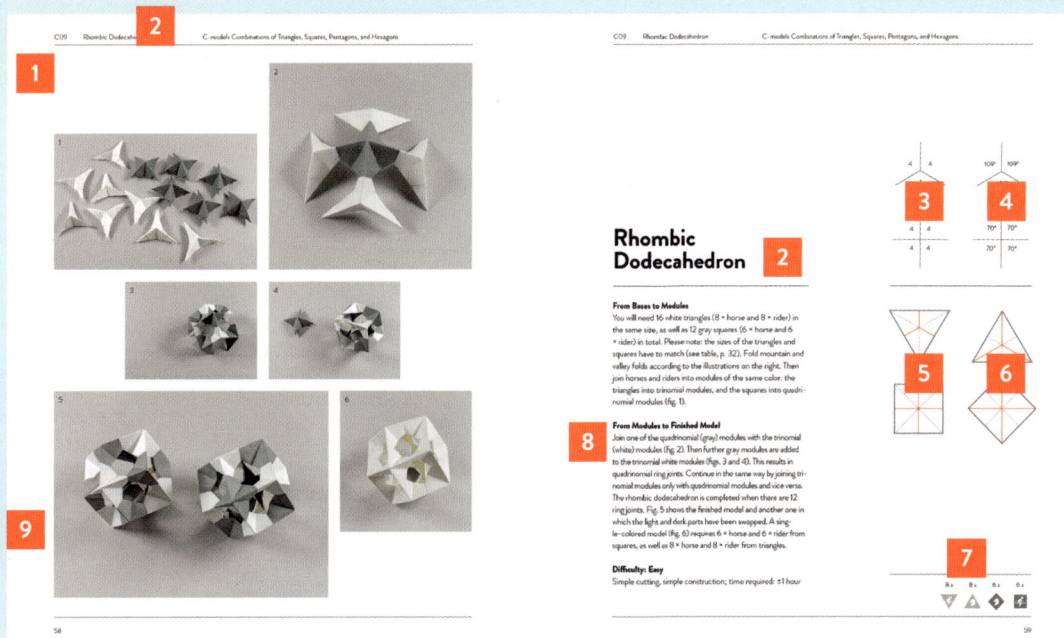

1 Model number
Letter: group of models (according to technical specification; e.g., models from triangles, squares, pentagons, and hexagons)
Number: running model number

2 Model name / technical specification
Example: C09 rhombic dodecahedron, belongs to the C group of models

3 Schematic illustration (module / ring joint)
Module and neighboring ring joints for the example of the rhombic dodecahedron (see p. 58): Each of the tri- and quadrinomial modules is bordered in this model only by quadrinomial ring joints.

4 Angles in the modules
Example rhombic dodecahedron: A trinomial module is bordered by ring joints at an angle of ±109°; a quadrinomial module is bordered by ring joints at an angle of ±70°.

5 Rider bases (schematic illustration)
Rider bases always rest on horse bases and are visible as the upper part of the module.

6 Horse bases (schematic illustration)
Horse bases are covered by rider bases and form the lower part of the module.

7 Pictograms
Rider is always on the top of a module.
Horse is always at the bottom of a module.
Centaurs are cut and put together.

8 Step-by-step instructions
Follow these as explained for the rhombic dodecahedron (see p. 59).

9 Step-by-step construction
Put the model together in sequence, as illustrated by the accompanying photographs (see p. 59).

Note on 5–7: Scaled images of the bases of horse and rider can be downloaded on a scale of 1:1 as templates at **www.schifferbooks.com/polyhedra**. As long as a model requires only one type of modules with horse-and-rider sheets of the same size, it can be constructed in any size.

Folded Polyhedra

All polyhedra models were executed using white paper. This allows the light to be reflected on the faces, and subtle shades of gray allow the polyhedron to emerge clearly in its three-dimensional shape. Additionally, strongly colored reflections from the immediate environment can play on the surfaces, which is revealed most beautifully in the changing light throughout the day when a model is placed near a window.

In addition, some models have instructions for building them in two or three colors, and each step has been photographed. Making models in several colors is especially suited as an introduction to the way of working and constructing used here. Where there is a joint between two different-colored modules, the joint becomes clearly visible.

When making your first polyhedra, the colored models make orientation easy. With these, you can gain some initial experience in how to understand the approach used. By following the details of the step-by-step instructions, you learn how to apply the instructions more independently.

In the central part of this book, the models have been illustrated using only white paper; the experience previously gained when making multicolored models then becomes a valuable help. This is the reason why I advise following the sequence given when building your own models: from the easy to the difficult, and from the simple to the complex. In the section "Extensions of Polyhedra Forms" (see p. 120 and following), the different colors signify different aspects of construction, depending on the type of fold.

Horse-and-Rider Function, Ring Joint

The principle of horse-and-rider function is the same across all models (i.e., the way modules are put together in a ring joint and the resulting rhythmical construction). However, these principles are continually extended and varied from one model to the next to conform to the geometrical conditions of construction required by each model in question. Models with comparable conditions of construction have been grouped together, usually forming a separate chapter.

A-models: Models that are easy to accomplish and that require only square sheets of paper as bases. Differences result from the number of modules, as well as from the omission of certain valley folds. They give you your first three-dimensional shapes. The differences between the models are interesting from a geometrical point of view.

All models (as closed polyhedra forms) are executed in approximately the same volume. The measurements of the individual parts are dependent on this. Sizes to scale of the individual bases can be downloaded as 1:1 templates (see p. 23) or found in the table on p. 32. As long as a model requires only one type of module with horse-and-rider sheets of the same size, it can be executed in any size.

B-models: Models that are similarly easy to accomplish and are based on triangles or pentagons. Here, too, the number of modules, and occasionally the omission of valley folds, leads to different three-dimensional shapes with certain similar characteristics.

C-models: First combinations of triangles with squares, triangles with pentagons, and pentagons with hexagons lead to further three-dimensional forms.

D-models: To achieve the desired measurements of the individual polyhedra forms, occasionally edge lengths are required that can be obtained only when horse and rider consist of bases of different sizes.

E-models: Large amounts of angle surplus (see p. 74) make additional mountain folds and infoldings necessary.

F-models: The necessary angles in a model have to be made to fit the angles possible to obtain in a module. A particular solution is needed here: both bases are cut/slit from one side so that they can be joined. Each base participates both in horse-and-rider functions. In reference to the horse/man beings in Greek mythology, these are called "centaurs" in *Folding Polyhedra*.

G-models: Particularly elaborate centaurs result from models whose sheets have two or three cuts/slits. It is better to attempt these models after having successfully accomplished the F-models.

H-models: The particularly small overlapping areas in the joints require additional supports that are fitted between horse and rider and serve to lend stability across modules (between two neighboring modules).

I-models: A little geometrical play: even icosahedra and dodecahedra can be approached and constructed as chiral models. The techniques previously introduced are applied here.

With each step, the challenges in constructing the individual models rise. In many cases, glue can be helpful. Especially G- and H-models pose great challenges both technically and because of the high number of necessary parts. It is thus advisable to always start with the simplest models and then approach the next challenges with growing experience. With particularly small parts, it can be an effective help to build the model from larger-sized bases. Doubling the edge length should be sufficient.

A-models
from Squares

The first three models have one thing in common: their base form is the square. However, they differ in the number of required parts and the number of mountain folds. These two aspects are related: the fewer mountain folds there are, the more parts are needed. Across the different polyhedra forms, these differences are revealed clearly. And the similarities emerge just as quickly: all three models have the same number of trinomial ring joints (eight triangles).

From Base to Module

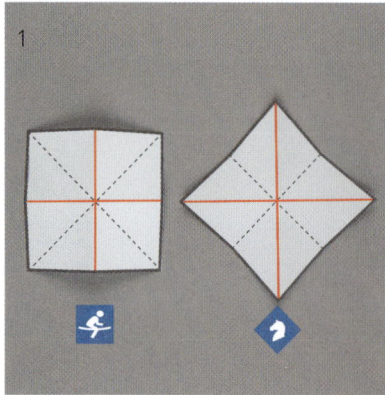

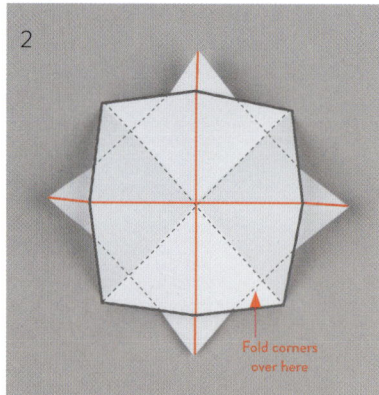

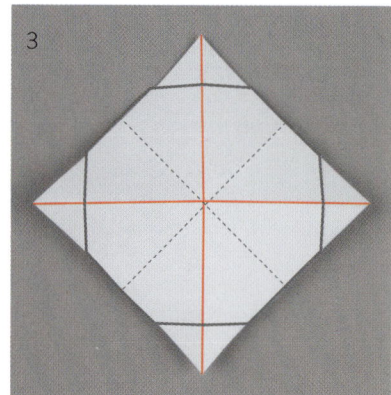

From Module to Model

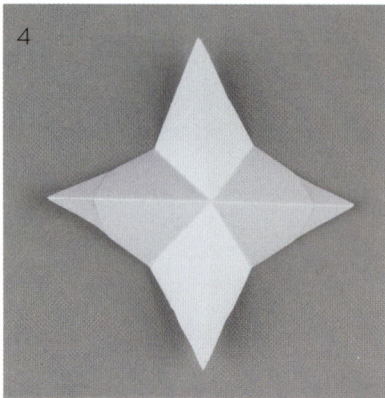

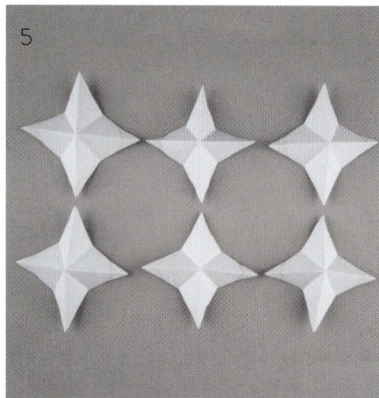

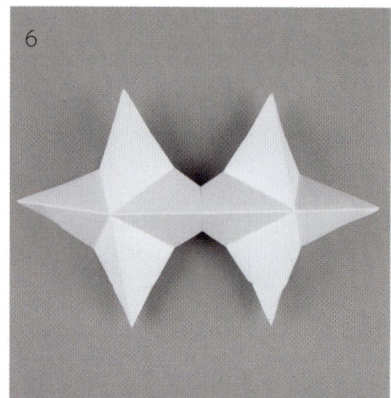

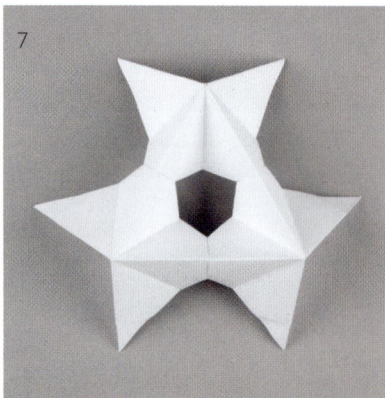

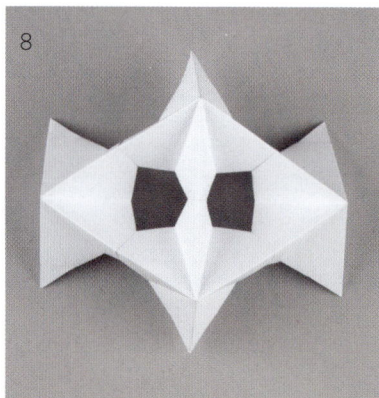

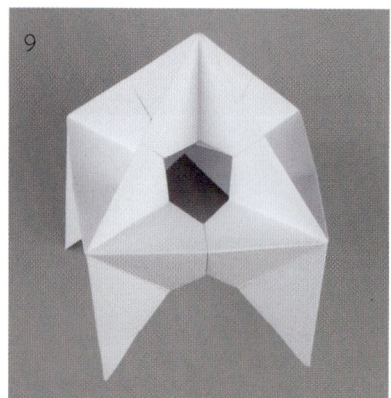

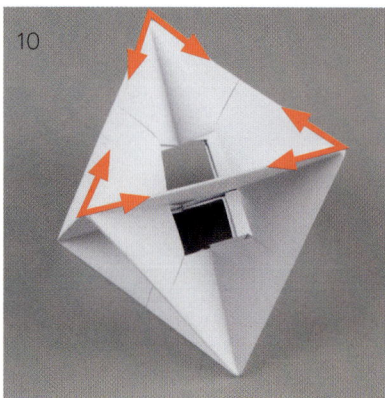

3	3		60°	60°
3	3		60°	60°

Octahedron

From Base to Module

You will need a total of 12 squares of equal size: (6 × horse and 6 × rider each). Fold the mountain and valley folds according to the illustration on the right. Then join the horse and rider by folding the rider one corners (fig. 3) to create one module (fig. 4). Create 6 modules (fig. 5).

From Module to Model

Connect two modules together (fig. 6). Add the third, all three modules are put together in a trinomial ring joint (fig. 7). Proceed in the same way, adding three further modules one after the other, so that further trinomial ring joints result: there are a total of eight trinomial ring joints in the completed model (fig. 10). The octahedron can also be built in two or three colors. In a two-colored model, the ring joints can be of the same color or of two colors (figs. 11).

Difficulty: Easy

Simple cutting, simple construction; time required: ±45 minutes

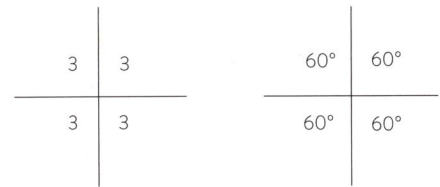

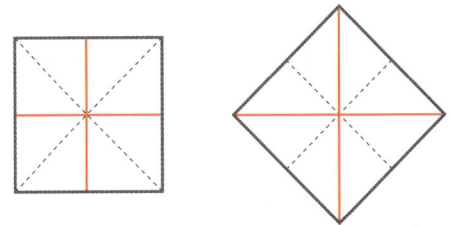

6 x

6 x

Hints and Details:

In the chapter "Practical Considerations" (see p. 22 and following), each step is explained in detail.

Figures for the sizes of bases (here, squares) for each form can be found in the table on p. 32.

From Base Area to Model

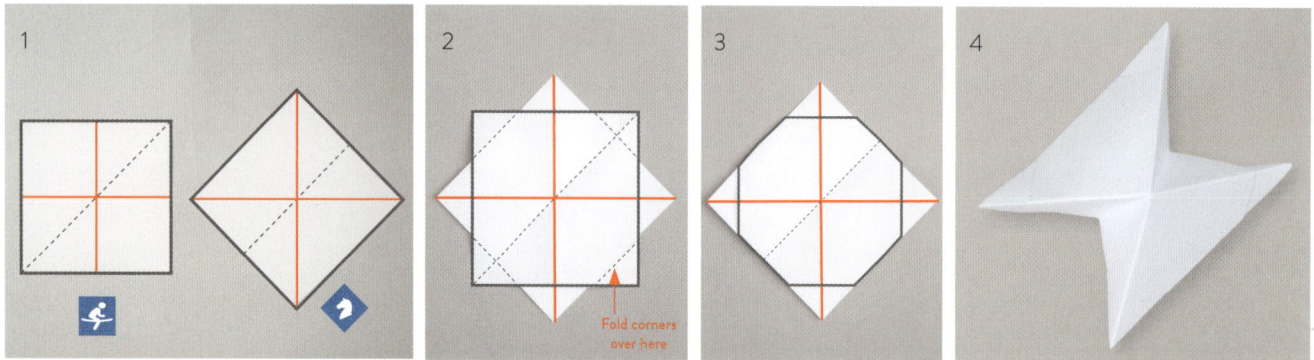

2 Fold corners over here

From Module to Model

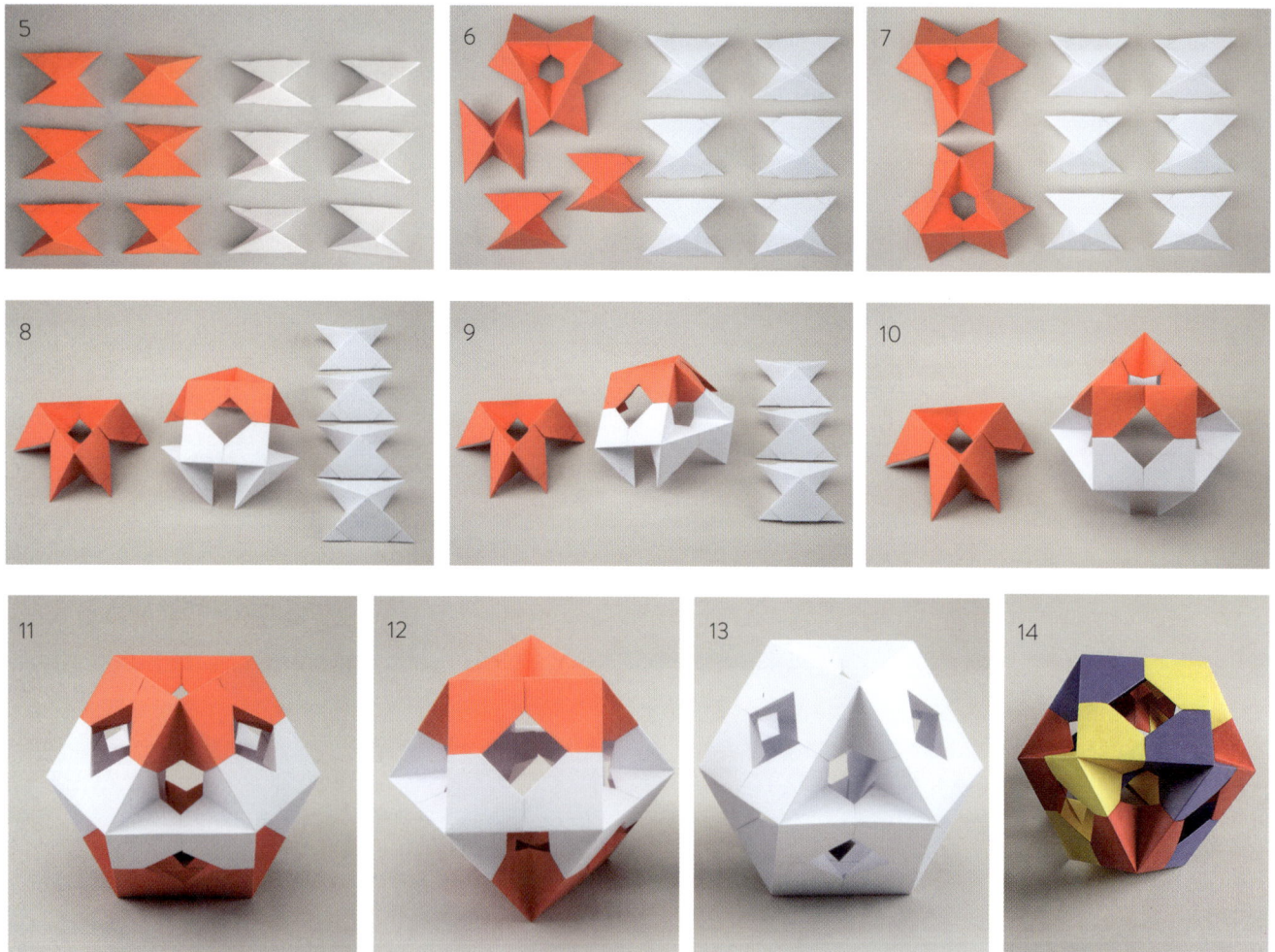

Cubo-octahedron

4	3		90°	60°
3	4		60°	90°

12 x   12x

From Base Area to Model
You will need a total of 24 squares of equal size: 12 in red and 12 in white (6 × horse and 6 × rider each). Fold the mountain and valley folds according to the illustration on the right. Then join horse and rider into modules (fig. 1-4). Create all 12 modules (fig. 5).

From Module to Model
Connect three red modules each together to create trinomial ring joints (fig. 5). Join the white modules with the red modules as shown in the picture. Repeat to create a second trinomial ring joint module (fig. 7). Combine red and two trinomial joint white modules to form quadrinomial ring joints (fig. 8). Add the other three white modules (fig. 9-10). The white modules form a kind of angular band around the completed model, and the red trinomial modules resemble polar caps (fig. 11). Alternatively, the model can be constructed in three colors. Each of the trinomial ring joints is then three colored, in the quadrinomial ones two colors, then alternate (fig. 14). Fig. 13 shows the model in one color. Connect the red trinomial module to complete the model.

Difficulty: Easy
Simple cutting, simple construction; time required: ±1 hour

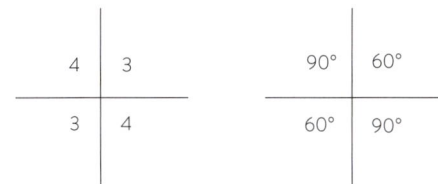

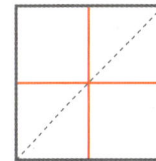

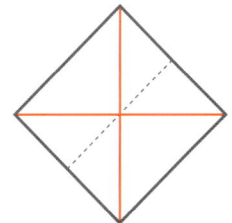

What are horse and rider?
The folded models all start from modules. Each module consists of two superseding paper bases. The upper is called the "rider" (see fig. left); the lower, the "horse" (see fig. right).

From Base Area to Module

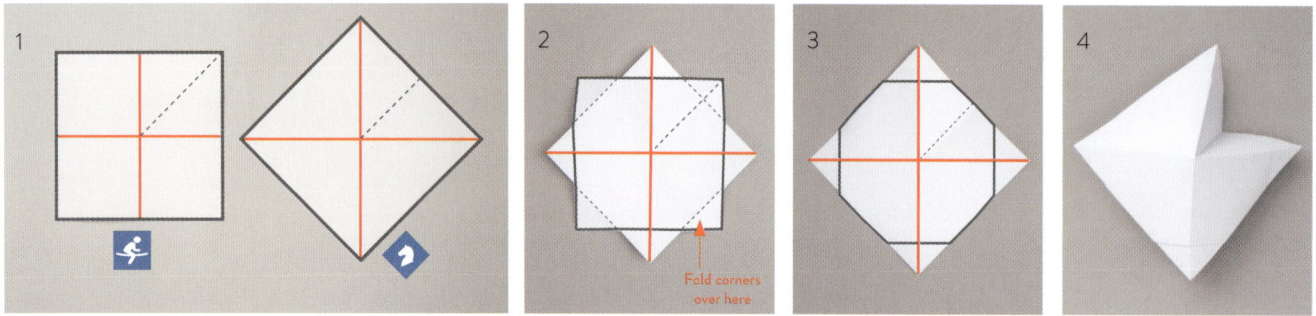

Fold corners over here

From Module to Model

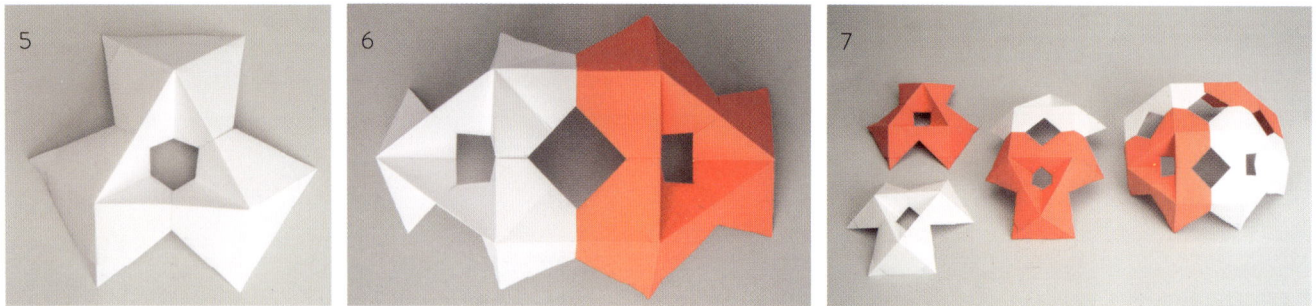

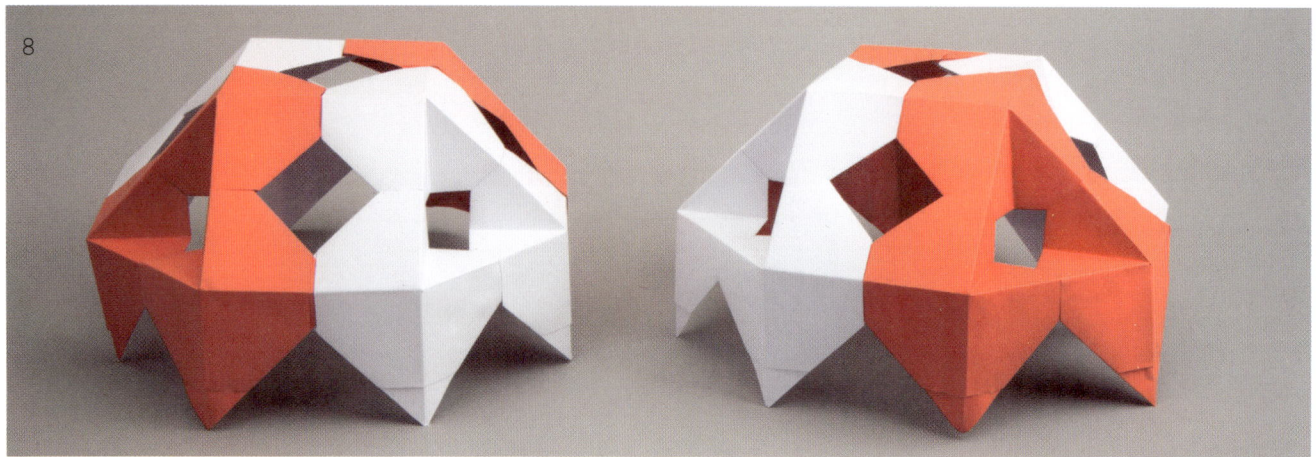

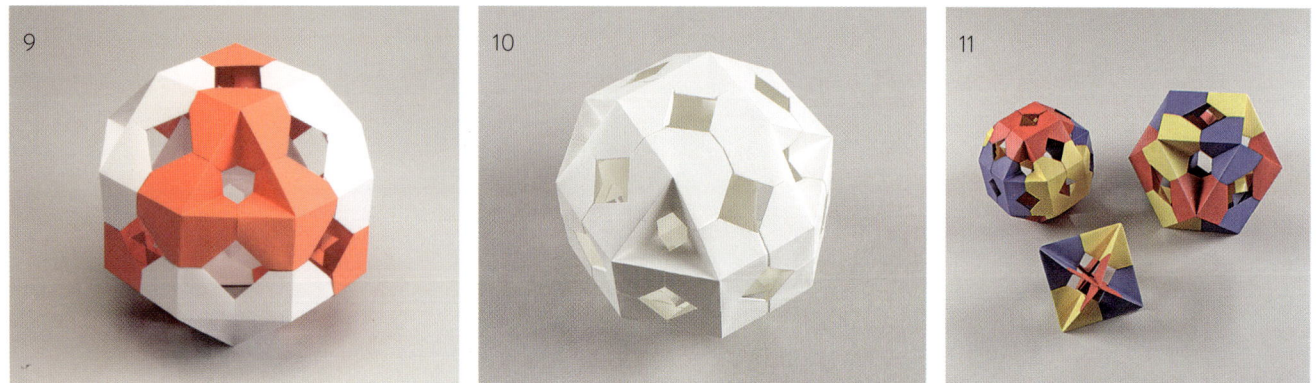

4	3		90°	60°
4	4		90°	90°

Rhombic Cubo-octahedron

First Steps
You will need a total of 48 squares of equal size: 24 in red and 24 in white (12 × horse and 12 × rider each). Fold the mountain and valley folds according to the illustrations on the right. Then join horses and riders into modules (fig. 1–4). Then create all 24 modules.

Constructing the Model
Combine three modules to form a trinomial ring (fig. 5). Create 8 of the same colored trinomial ring forms from the modules. Connect the red and white rings; this results in new, binomial ring joints (fig. 6). Continue in the same way by adding further trinomial rings so that a three-dimensional pattern, similar to a chessboard, is created (fig. 7). Form two halves of the model (fig. 8), then connect to complete the model (fig. 9). Fig. 10 shows the model in one color and Fig. 11 shows the model in three colors.

Difficulty: Easy
Simple cutting, simple construction; time required: ±75 minutes

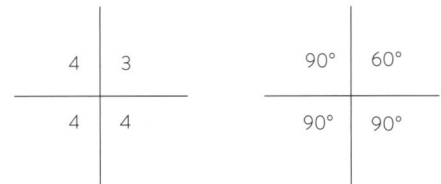

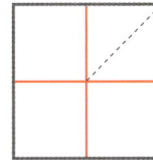

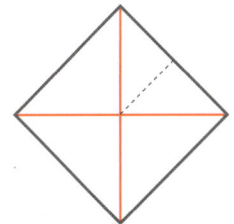

24 x

24 x

Hint:
Always create modules in one color (e.g., white horse + white rider = white module).

B-models
from Triangles or Pentagons

In contrast to the preceding chapter on folded polyhedra, this chapter introduces the first models from triangles. When you go through all possible valley folds, a further model becomes possible.

If a valley fold is omitted, three different models can be made with the resulting module—depending on whether you create tri-, quadri-, or pentanomial ring joints. The similarities between the three models are obvious. One model based on pentagons supplements this sequence.

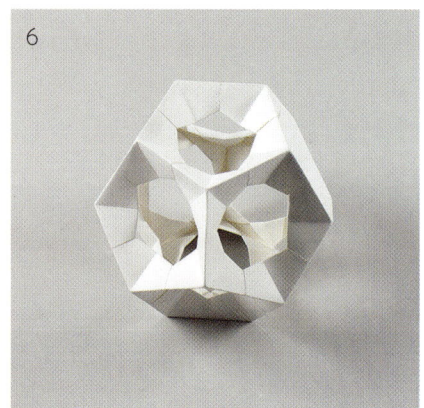

Dodecahedron

First Steps
A total of 40 triangles of equal size are needed: 20 in gray and 20 in white (10 × horse and 10 × rider each).

Constructing the Model
Fold mountain and valley folds according to the illustrations on the right, and join horse and rider into modules. Then join the gray and white modules, respectively, into a fivefold ring each (fig. 1). Add gray modules to the white ring and then add further white modules to the resulting gray ring. Finally, add the fivefold gray ring (figs. 2–4). The completed model has a dark and a light polar cap; in between, there is a light and dark band running all the way around (fig. 5). Figure 6 shows the dodecahedron in one color.

Difficulty: Easy
Simple cutting, simple construction; time required: ±1 hour

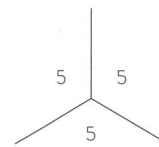

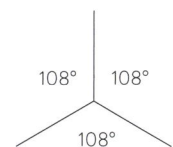

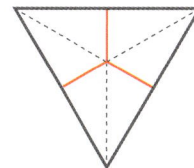

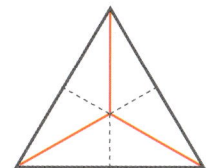

5   5

108°   108°

5

108°

10 x   10 x   10 x   10 x

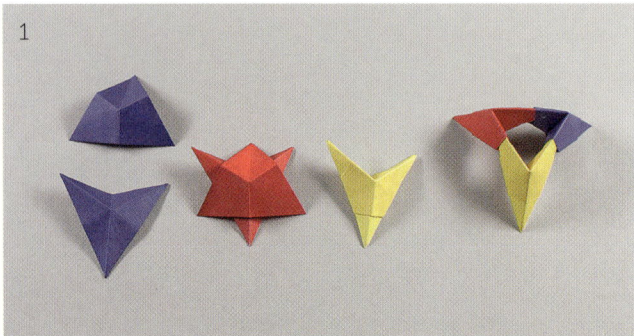

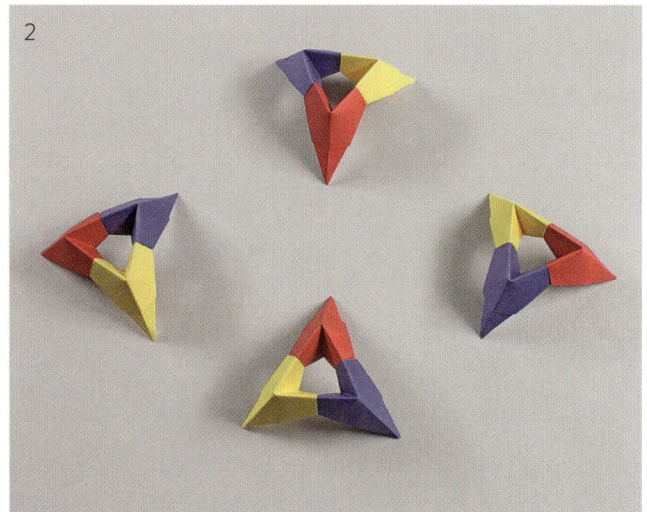

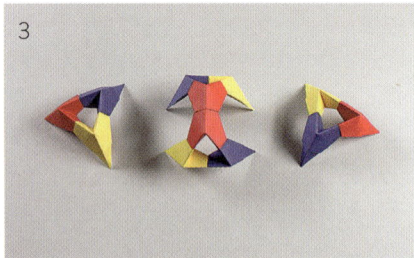

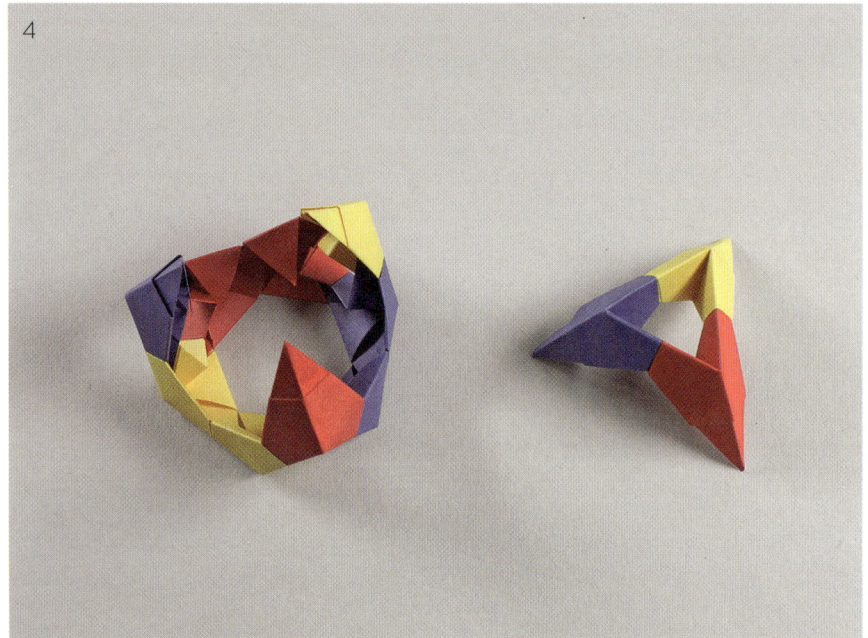

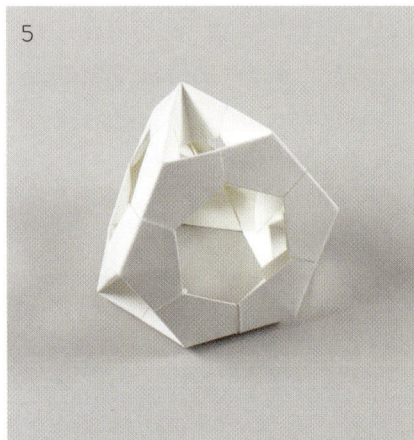

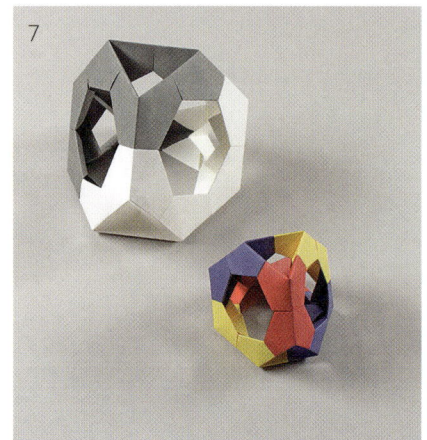

Truncated Tetrahedron

First Steps

A total of 24 triangles of equal size are needed: 8 in yellow, 8 in red, and 8 in blue (4 × horse and 4 × rider each). Fold the mountain and valley folds according to the illustrations on the right. Then join horse and rider into modules of the same color and create trinomial ring joints in three colors (fig. 1, *from left to right*; fig. 2).

Constructing the Model

Take two ring joints and join them so that two modules of the same color meet (fig. 3), then add the third ring joint (fig. 4). The fourth ring joint completes the model (fig. 6). You can also make the truncated tetrahedron in two colors (fig. 7). In this case, you need 6 × horse and 6 × rider in gray and the same number in white. A model in one color (fig. 5) requires 12 × horse and 12 × rider of the same color and size.

Difficulty: Easy

Simple cutting, simple construction; time required: ±75 minutes

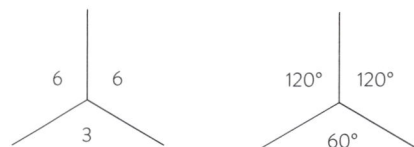

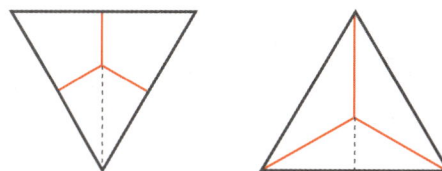

Why are some models multicolored? The places where the individual modules meet are easy to see in colored models. Constructing the model in colors thus contributes to a better overview. Moreover, colors help to highlight special symmetries.

| 4 x | 4 x | 4 x | 4 x | 4 x | 4 x |

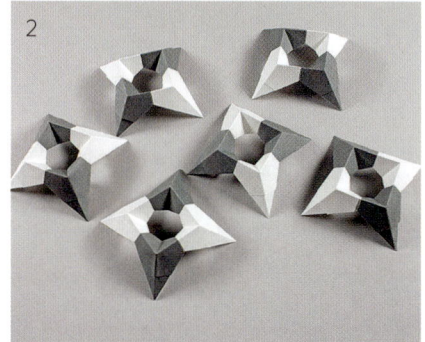

Truncated Octahedron

First Steps

You will need a total of 24 gray triangles of equal size, consisting of 12 × horse and 12 × rider, and the same number in white. Fold the mountain and valley folds according to the illustrations on the right. Join horse and rider into modules of the same color. Join all fourfold ring joints, alternating the two colors (fig. 1, *from left to right*, as well as fig. 2).

Completing the Model

Combine two ring joints so that two modules of different colors meet (fig. 3). Add all remaining ring joints in sequence (figs. 3 and 4); the sixth ring joint completes the model (fig. 5). You can also make the truncated octahedron in two different versions, using three colors (fig. 7). For each model you need 8 × horse and 8 × rider each in yellow, red, and blue. A model in one color only (fig. 6) requires 24 × horse and 24 × rider in the same color and size.

Difficulty: Easy

Simple cutting, simple construction; time required: ±75 minutes

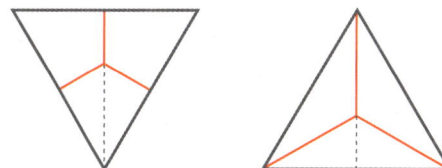

1

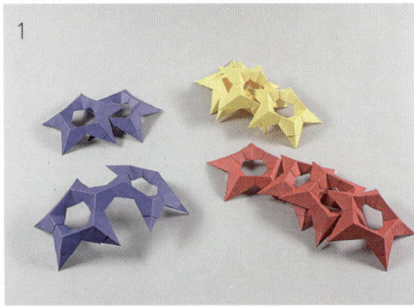

2

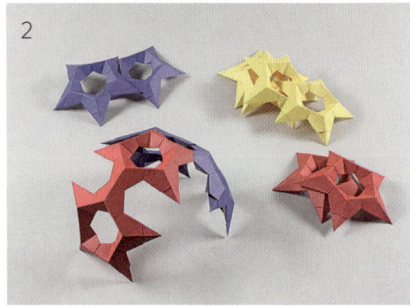

3

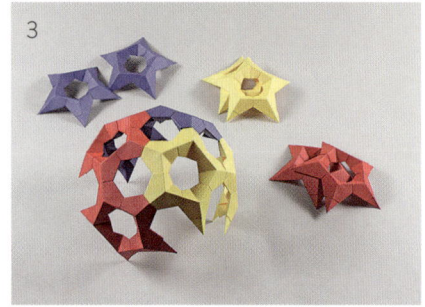

4

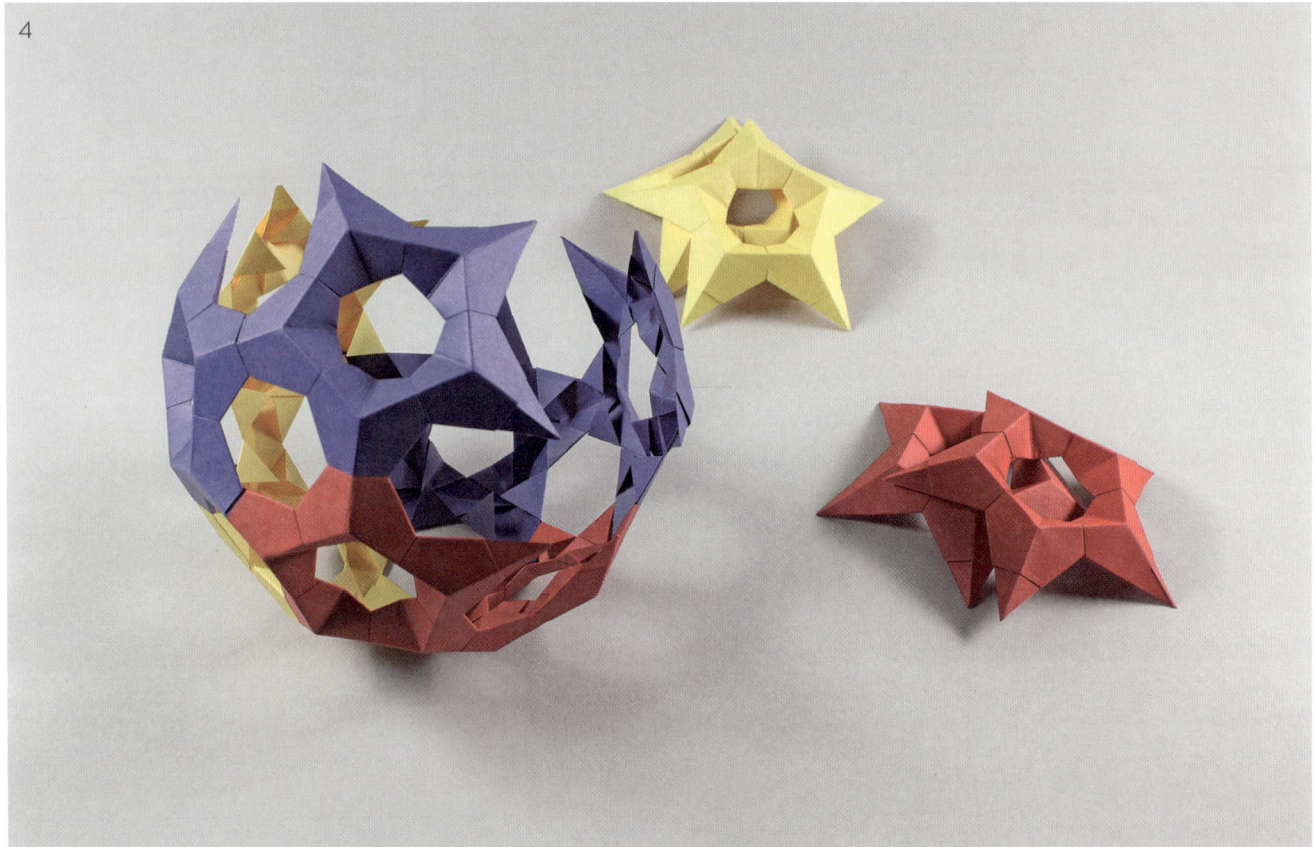

5

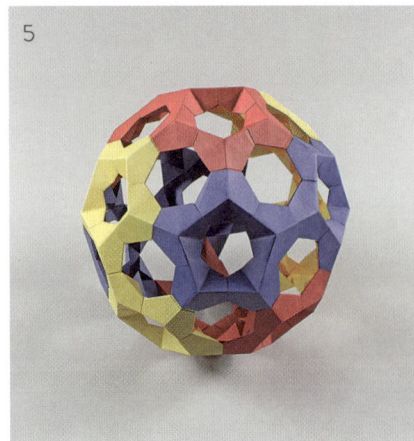

6

7

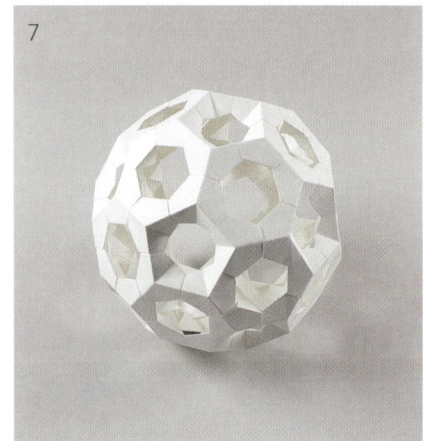

Truncated Icosahedron

First Steps

For a model in three colors, a total of 120 triangles of the same size are needed: 40 in yellow, 40 in red, and 40 in blue (20 × horse and 20 × rider each). Fold mountain and valley folds according to the illustrations on the right. Subsequently, join horse and rider into modules of one color (see preceding models B05 and B06, p. 48 and following).

Completing the Model

Create fivefold ring joints and join two of these each in the same color into fivefold ring joints (figs. 1 and 2). Then add further double ring joints (figs. 3–5) until the model is complete.

You can also make the truncated icosahedron in two colors (fig. 6). In this case you need 30 × horse and 30 × rider in gray and the same number of white bases. A model in only one color (fig. 7) requires 60 × horse and 60 × rider in one color and size.

Difficulty: Easy

Simple cutting, simple construction; time required: ±3 hours

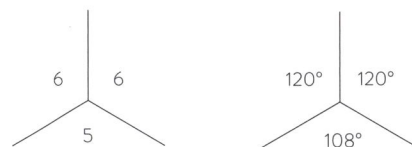

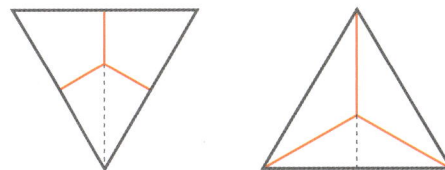

| 20 x | 20 x | 20 x | 20 x | 20 x | 20 x |

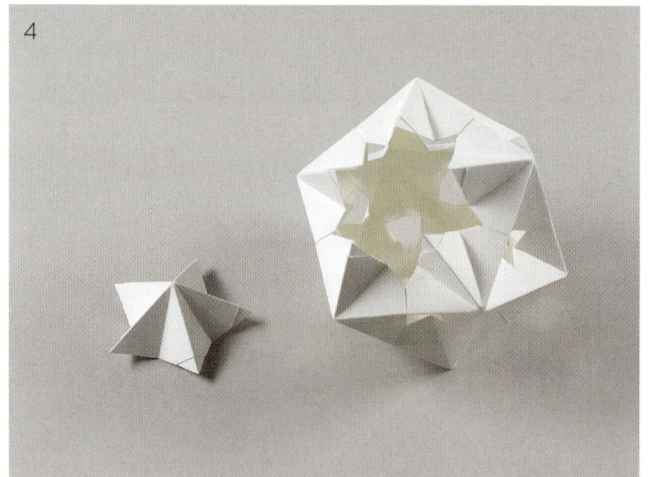

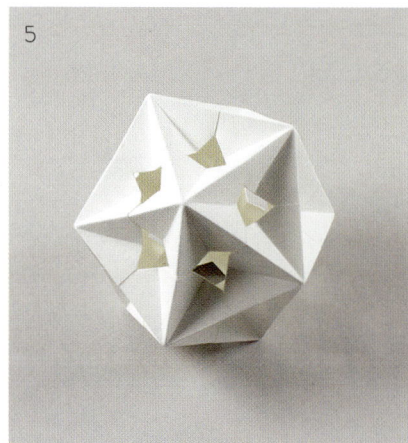

Icosahedron

From Bases to Modules

You will need a total of 24 pentagons, consisting of 12 × horse and 12 × rider in the same size. Fold mountain and valley folds according to the illustrations on the right. Then join horse and rider into modules (fig. 1).

From Module to Finished Model

Join the first three modules into a trinomial ring joint (fig. 2). Then join the remaining modules one after the other into further trinomial ring joints (figs. 3 and 4). In an icosahedron (= 20 surfaces) there are eventually exactly 20 trinomial ring joints. Fig. 5 shows the completed icosahedron.

Difficulty: Easy

Simple cutting, simple construction; time required: ±1 hour

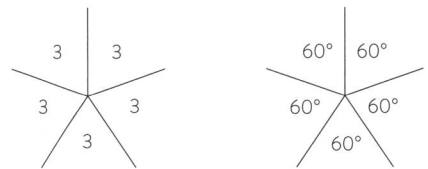

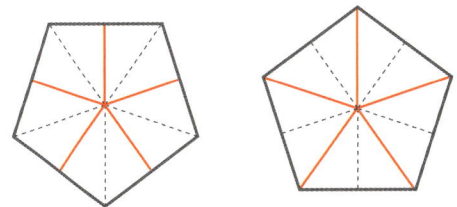

12 x   12 x

C-models
Combinations of Triangles, Squares, Pentagons, and Hexagons

So far, models have been presented consisting exclusively of parts that are triangular, square, or pentagonal in shape. Now we will combine different base shapes in one model. To begin, triangles are combined with squares, then triangles with pentagons. In both models, rhombic ring joints can be found (rhombic dodecahedron and rhombic hexecontahedron). And finally, pentagons are joined with a new form—hexagons—into the triakis icosahedron. This polyhedron shape is familiar from the radar domes that can be seen in many places.

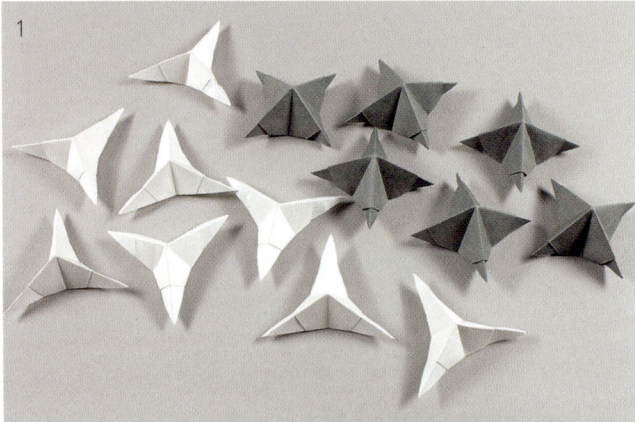

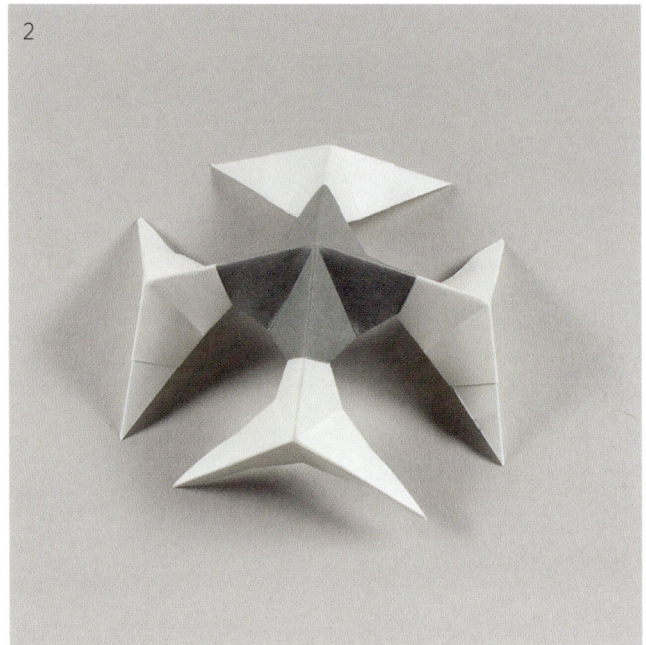

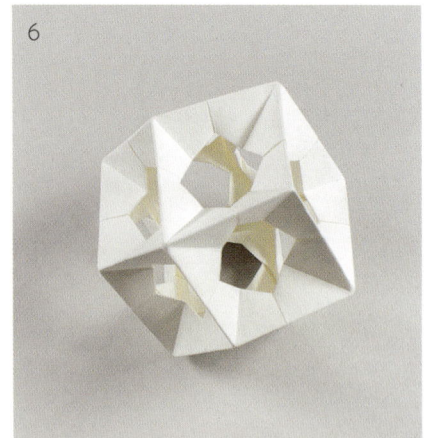

Rhombic Dodecahedron

From Bases to Modules

You will need 16 white triangles (8 × horse and 8 × rider) in the same size, as well as 12 gray squares (6 × horse and 6 × rider) in total. Please note: the sizes of the triangles and squares have to match (see table, p. 32). Fold mountain and valley folds according to the illustrations on the right. Then join horses and riders into modules of the same color: the triangles into trinomial modules, and the squares into quadrinomial modules (fig. 1).

From Modules to Finished Model

Join one of the quadrinomial (gray) modules with the trinomial (white) modules (fig. 2). Then further gray modules are added to the trinomial white modules (figs. 3 and 4). This results in quadrinomial ring joints. Continue in the same way by joining trinomial modules only with quadrinomial modules and vice versa. The rhombic dodecahedron is completed when there are 12 ring joints. Fig. 5 shows the finished model and another one in which the light and dark parts have been swapped. A single-colored model (fig. 6) requires 6 × horse and 6 × rider from squares, as well as 8 × horse and 8 × rider from triangles.

Difficulty: Easy

Simple cutting, simple construction; time required: ±1 hour

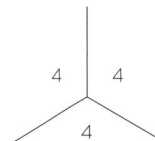

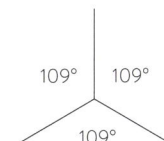

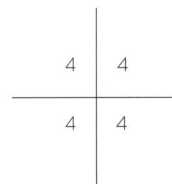

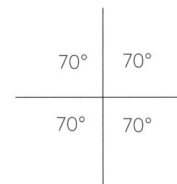

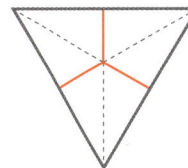

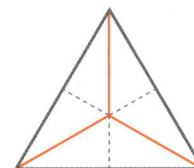

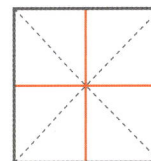

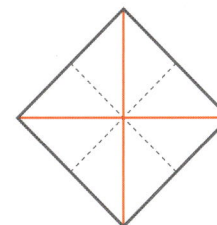

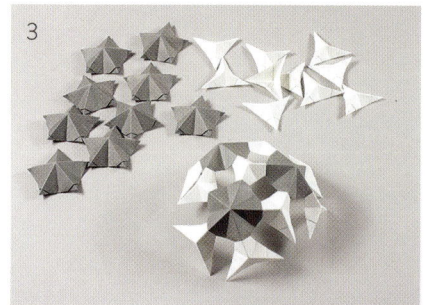

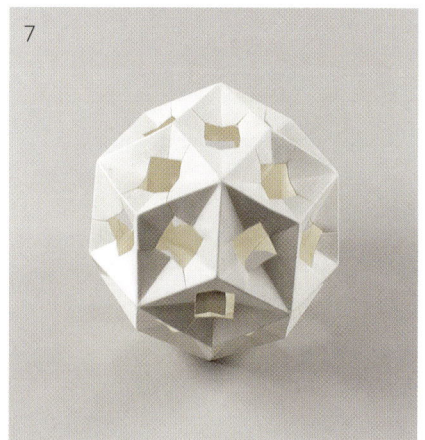

Rhombic Tricontahedron

From Bases to Modules

You will need a total of 40 white triangles (20 × horse and 20 × rider) as well as 24 gray pentagons (12 × horse and 12 × rider). Please note: the sizes of the triangles and pentagons have to correspond (see table, p. 32). Fold mountain and valley folds according to the illustration on the right. Then join horse and rider into single-colored modules: the triangles into threefold modules and the pentagons into fivefold modules (fig. 1).

From Modules to Finished Model

Join one of the trinomial (white) modules with three pentanomial (gray) modules (fig. 2). Further trinomial white modules are added to pentanomial gray modules (fig. 3). This leads to four quadrinomial ring joints (figs. 4 and 5). Continue in this manner by joining trinomial modules exclusively with pentanomial modules and vice versa.

The rhombic tricontahedron is finished when 30 ring joints have been created (fig. 6). Fig. 7 shows a model in one color.

Difficulty: Medium

Cutting: first challenges; simple construction; time required: ±2 hours

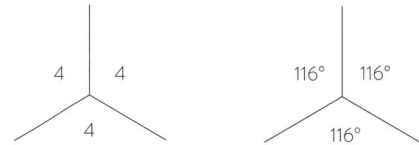

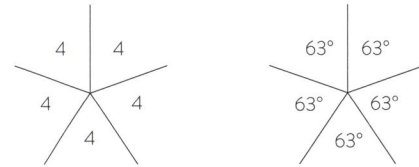

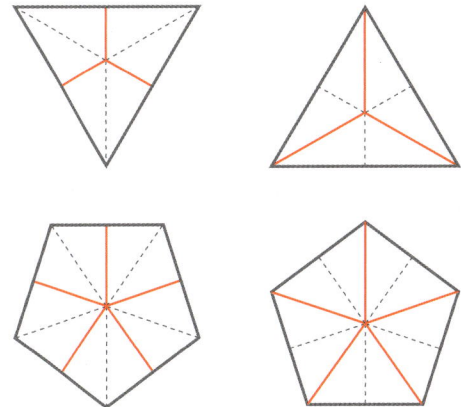

20 x   20 x   12 x   12 x

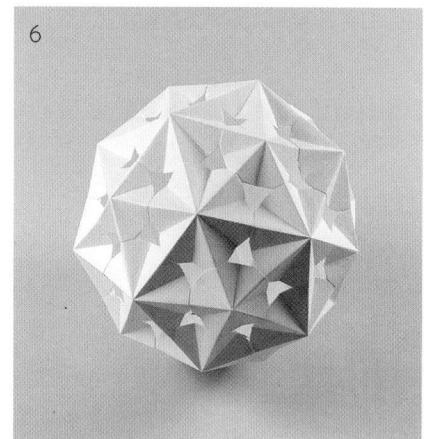

Triakis Icosahedron

From Bases to Modules

You will need a total of 24 pentagons of the same size (12 × horse and 12 × rider) as well as 40 hexagons (20 × horse and 20 × rider). Please note: the sizes of the hexagons and pentagons have to correspond (see table, p. 32). Fold mountain and valley folds according to the illustrations on the right. Then join horse and rider into modules: the pentagons into fivefold modules, the hexagons into sixfold modules (fig. 1, *top left*).

From Modules to Finished Model

Join one of the fivefold modules with five sixfold modules (fig. 1, *bottom right*). Further fivefold modules are joined with the sixfold modules (fig. 2). In the same manner, add further individual modules subsequently. Trinomial ring joints are the result (fig. 2).

Note the following principle: fivefold modules are joined only with sixfold modules, and sixfold modules alternatingly with five- and sixfold modules (figs. 3–5). The triakis icosahedron is finished (fig. 6) when 60 ring joints have been created.

Difficulty: Medium

Cutting: first challenges; simple construction; time required: ±3 hours

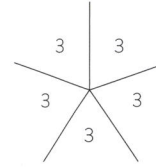

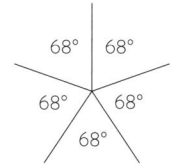

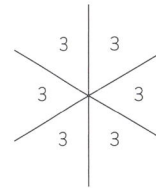

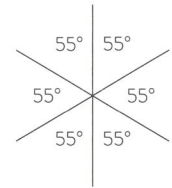

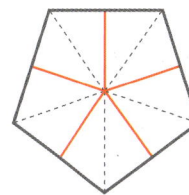

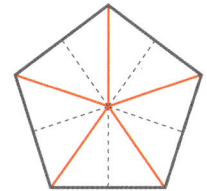

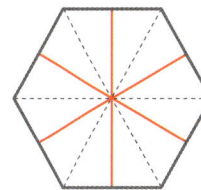

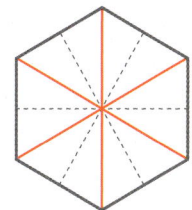

D-models
Horse and Rider in Different Sizes

When horse and rider of different sizes are joined into a module, the edge lengths in the model will change. A larger rider makes an edge longer; a larger horse makes it smaller. The deltoidal icositetrahedron and deltoidal hexecontahedron have many similarities: in these models, we encounter deltoid ring joints.

The pentagonal icositetrahedron and pentagonal hexecontahedron have irregular fivefold ring joints. The other thing that is new in this chapter is that these two forms can also be joined into a mirror ("chiral") variant.

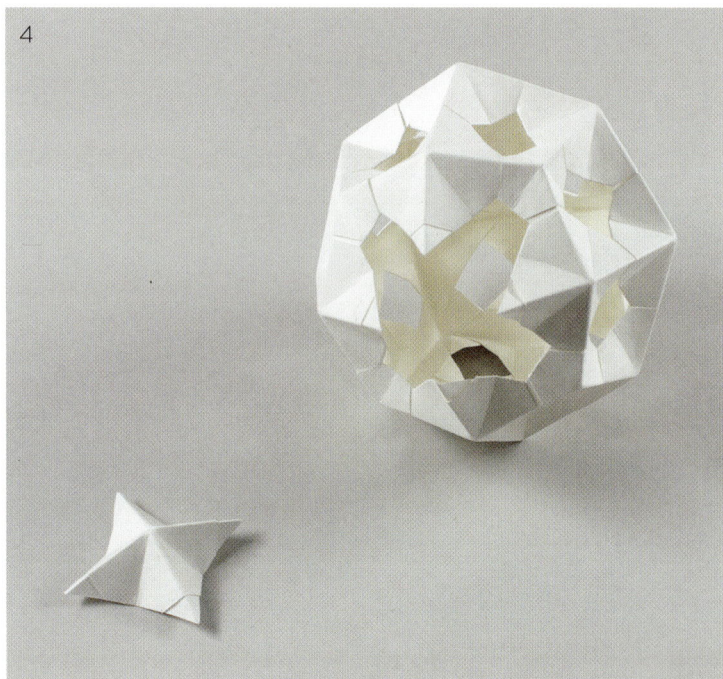

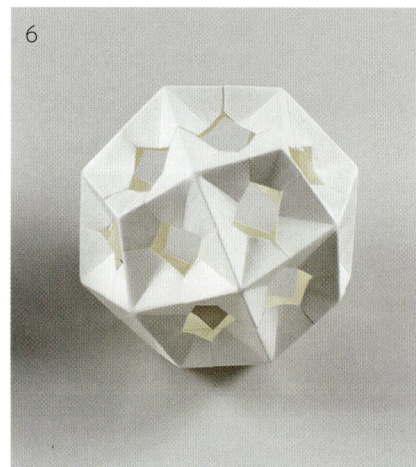

Deltoidal Icositetrahedron

From Bases to Modules

You will need a total of 16 triangles in different sizes, of which 8 × horse (bigger than rider) and 8 × rider (smaller than horse), as well as 36 squares (18 × horse and 18 × rider), are of the same size. Note the sizes listed in the table on p. 32. Fold the mountain and valley folds according to the illustrations on the right. Then join horse and rider into modules (fig. 1, *top*).

From Modules to Finished Model

Join a first trinomial module with three quadrinomial modules (fig. 1, *bottom*) and add further quadrinomial modules one after the other (fig. 2). This results in quadrinomial deltoidal ring joints.

Note the following principle: Trinomial modules are exclusively joined with quadrinomial modules. Some quadrinomial modules are joined alternatingly with tri- and quadrinomial modules, others only with quadrinomial modules (figs. 3–5). The deltoidal icositetrahedron is finished when 24 ring joints have been created (fig. 6).

Difficulty: Medium

Cutting: first challenges; construction: first challenges; time required: ±1½ hours

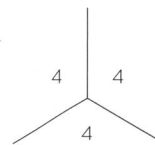

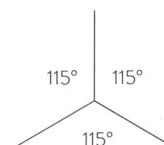

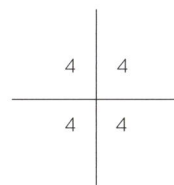

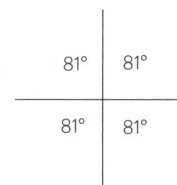

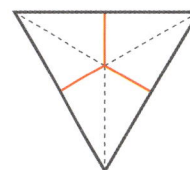

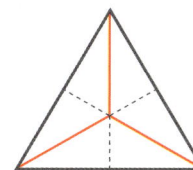

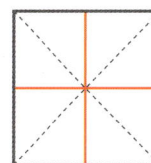

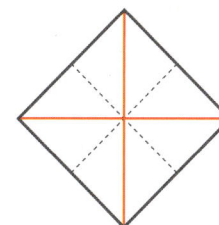

4 4	115° 115°
4	115°
4 4	81° 81°
4 4	81° 81°

8 x	8 x	18 x	18 x

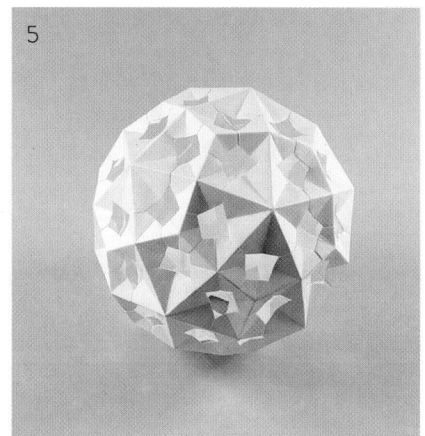

Deltoidal Hexecontahedron

From Bases to Modules
You will need a total of 40 triangles of the same size (20 × horse and 20 × rider), 60 squares of the same size (30 × horse and 30 × rider), and 24 pentagons of different sizes (12 × bigger rider and 12 × smaller horse). Note the sizes listed in the table on p. 32. Fold mountain and valley folds according to the illustrations on the right. Then join horse and rider into modules.

From Modules to Finished Model
Five quadrinomial modules are added to a first pentanomial module. Add trinomial modules between the quadrinomial modules (fig. 1) and complete the model by continuing in the same manner (figs. 2–5).
Note the following principle: Join each pentanomial module exclusively with quadrinomial modules, and each trinomial module only with quadrinomial modules. Each quadrinomial module is alternatively joined with trinomial and pentanomial modules.

Difficulty: Medium
Cutting: first challenges; simple construction; time required: ±4 hours

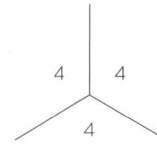

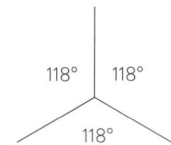

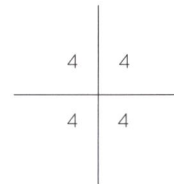

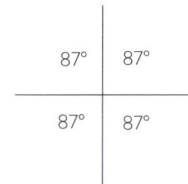

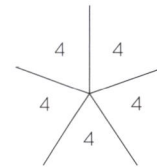

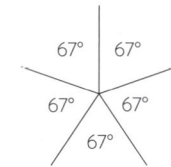

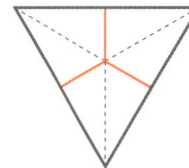

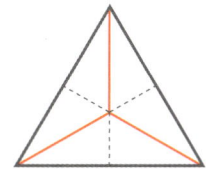

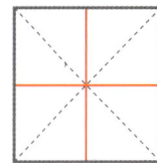

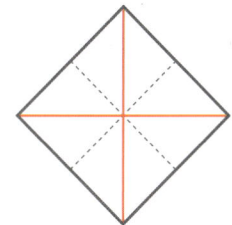

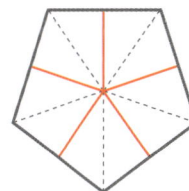

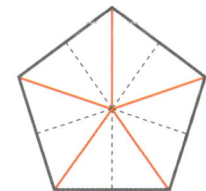

| 20 x | 20 x | 30 x | 30 x | 12 x | 12 x |

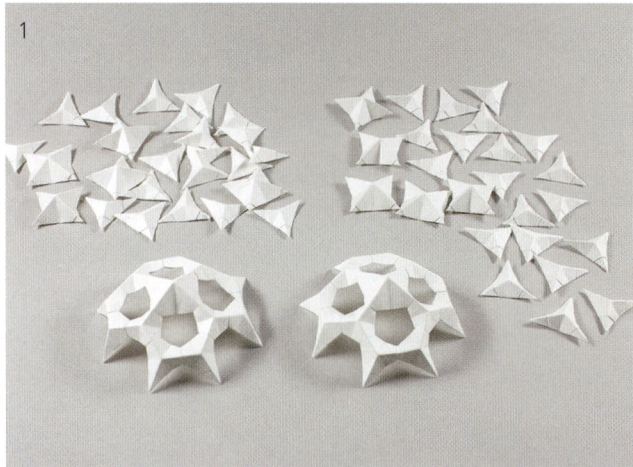

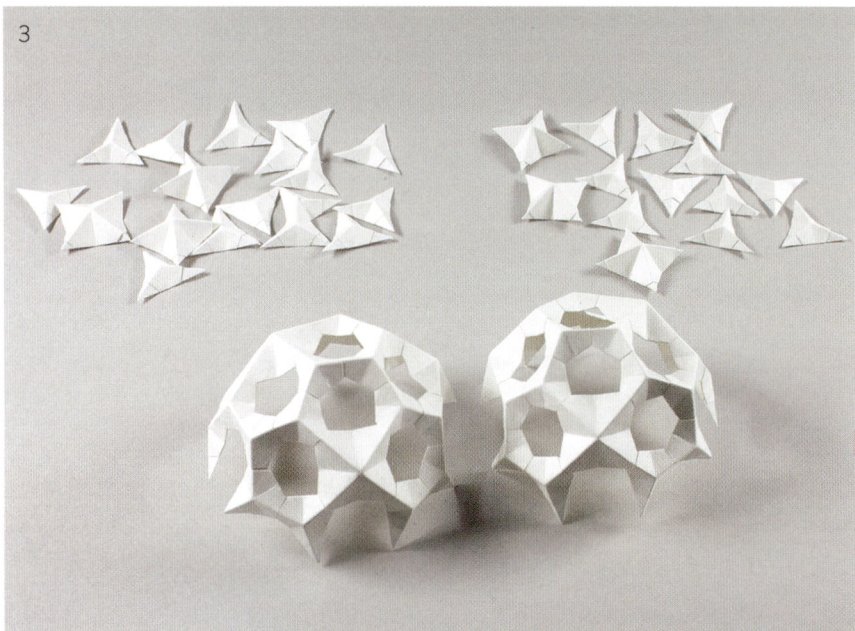

Pentagonal Icositetrahedron (Chiral)

From Bases to Modules

You will need a total of 64 triangles of the same size (32 × horse and 32 × rider) as well as 12 squares in different sizes (6 × larger horse and 6 × smaller rider). Note the sizes listed in the table on p. 32.

Fold the mountain and valley folds according to the illustrations on the right. Then join horse and rider into modules.

From Modules to Finished Model

Join one of the square modules on four sides with triangular modules. Then create a pentamerous ring joint on each side, using further trinomial modules (fig. 1). Continue in the same way by adding further pentamerous ring joints (figs. 2–4). Note the following principle: Each pentamerous ring joint consists of one square and four triangular modules.

Two models can be made of this folded polyhedron (chiral forms; fig. 5).

Difficulty: Medium

Cutting: first challenges, construction: first challenges; time required: ±1½ hours per model

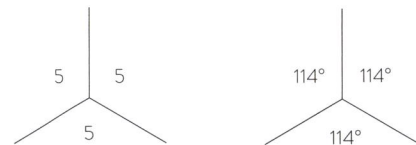

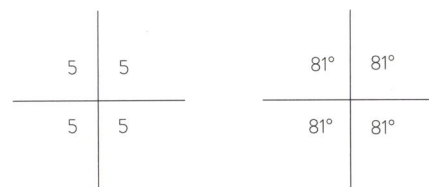

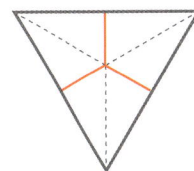

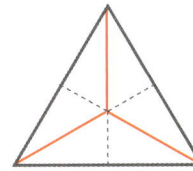

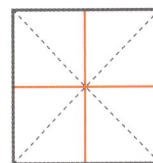

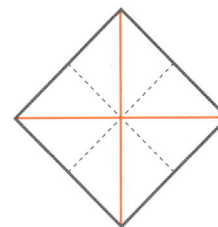

| 5 | 5 | 114° | 114° |
| 5 | | 114° | |

| 5 | 5 | 81° | 81° |
| 5 | 5 | 81° | 81° |

| | 32 x | 32 x | 6 x | 6 x |

1

2

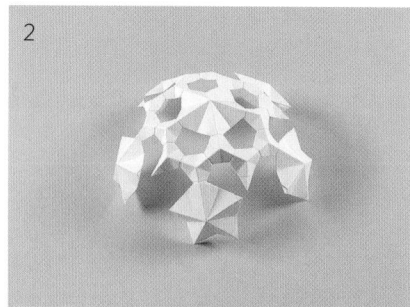

3

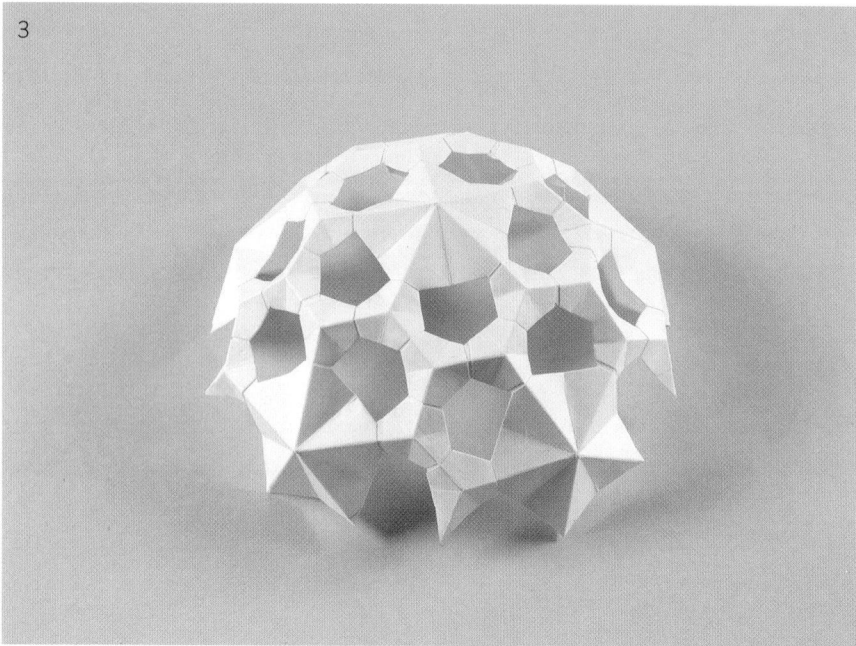

4

6

5

Pentagonal Hexecontahedron (Chiral)

From Bases to Modules

You will need a total of 160 triangles of the same size (80 × horse and 80 × rider) as well as 24 pentagons in different sizes (12 × larger horse and 12 × smaller rider). Note the sizes listed in the table on p. 32. Fold the mountain and valley folds according to the illustrations on the right. Then join horse and rider into modules.

From Modules to Finished Model

Join triangular modules on five sides with one of the pentagonal modules (fig. 1). Then create a pentamerous ring joint on each side, using further trinomial modules. Continue in the same way by adding further pentamerous ring joints (fig. 2 and figs. 3–5).

Note the following principle: each pentamerous ring joint consists of one pentagonal and four triangular modules. This folded polyhedron can be created in two different models (chiral forms; fig. 6).

Difficulty: Medium

Cutting: first challenges; construction: first challenges; time required: ±5½ hours per model

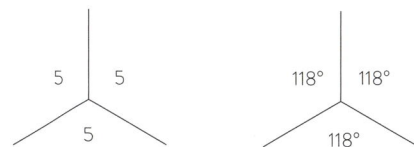

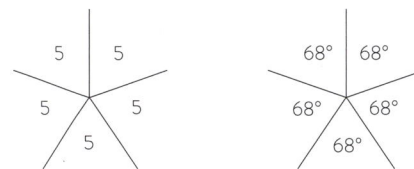

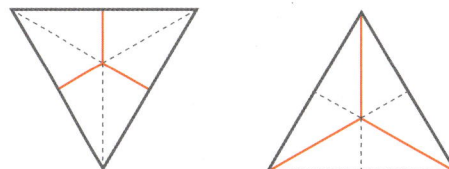

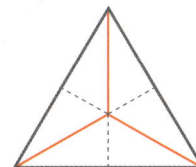

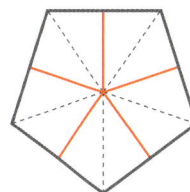

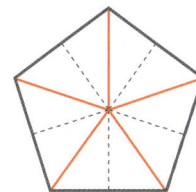

80 x   80 x   12 x   12 x

E-models
Additional Mountain Folds and Enveloping

The following two models consist of triangles and have one thing in common: a considerable proportion of module surface area is so large that it would be in the way when constructing the models. To understand the geometry of this problem, please read the explanations regarding the subject "mesh gap" in the chapter "Practical Considerations" (see p. 26). One practical solution is to fold excess surface area into the model by using additional mountain folds—and in the second model (a tetrahedron), to wrap or envelop them all the way to the inside of the model.

1

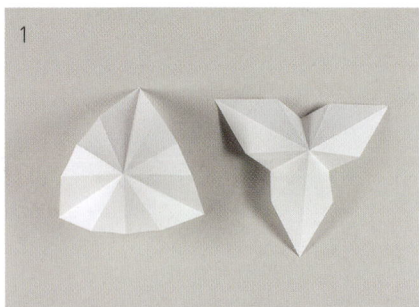

2

3

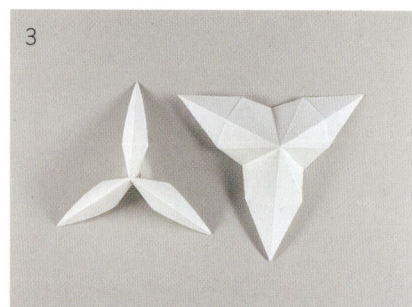

4

5

6

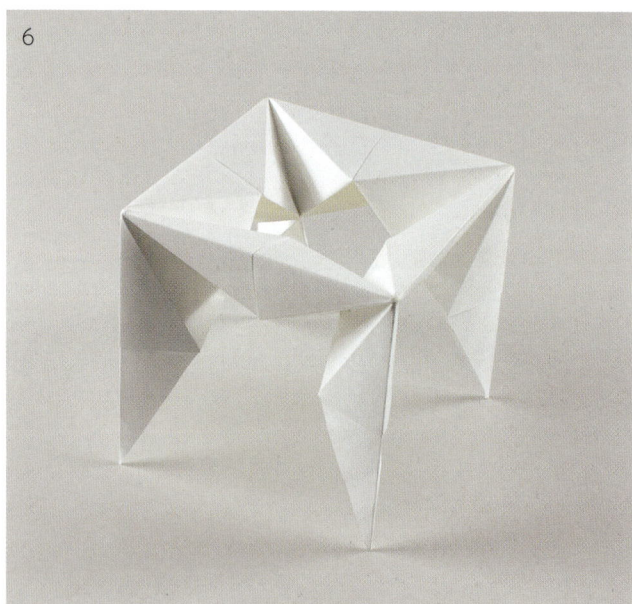

7

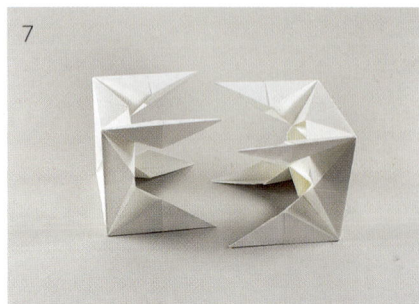

8

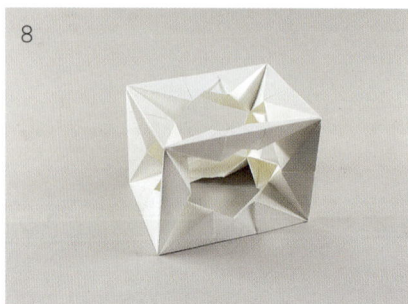

9

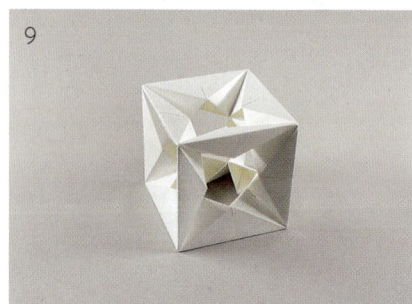

Cube

From Base to Module
You will need a total of 16 triangles in different sizes (8 × smaller horse and 8 × larger rider). Note the sizes given in the table on p. 32. Fold mountain and valley folds according to the illustrations on the right (fig. 1). Note the additional mountain folds. Join horse and rider into modules (figs. 2–3). Then join the centers of three valley folds each, using glue (alternatively sew them together; fig. 4).

From Module to Finished Model
Join one module on one side with other modules (fig. 5). Add two further modules, creating a first quadrinomial ring joint (first half of the cube; fig. 6). Build the other half of the cube in the same way and then put the two halves together (figs. 7–9).

Difficulty: Challenging
Particular challenges through additional folds and the necessity of gluing (or sewing); time required: ±1¼ hours

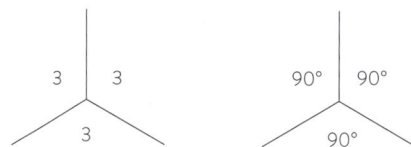

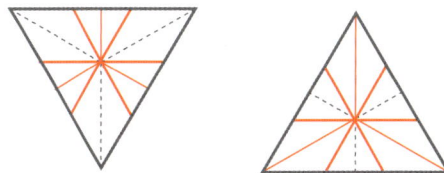

Note:
The mesh gap (i.e., the excess-angle area) in the cube and in the tetrahedron are particularly large. Paper clips facilitate the gluing of the three valley folds (fig. 4). In the cube, only a third of the total area is needed: out of 360°, only 3 × 90° = 270°. For the folded model, this means an angle excess of 30° on each side. It is not possible to fold this inward with a valley fold as before; two additional mountain folds are needed. Three valley folds each are then glued together.

8 x   8 x

1

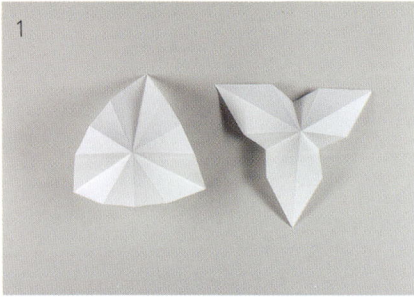

2

3

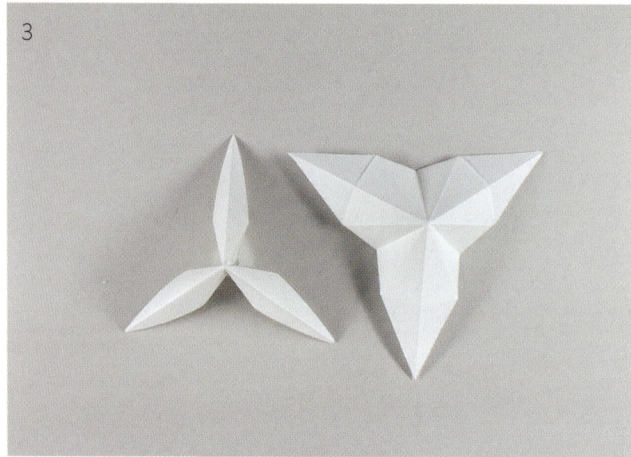

4

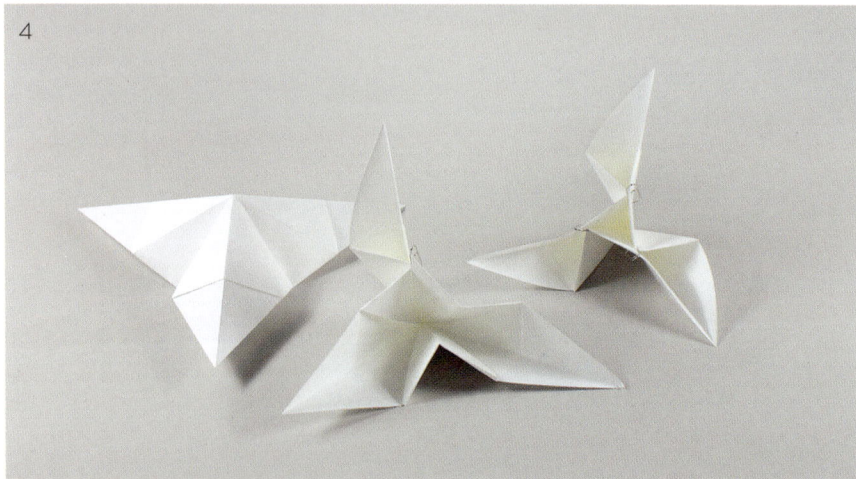

5

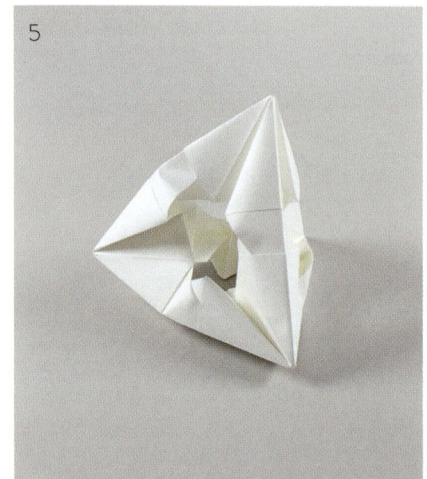

6

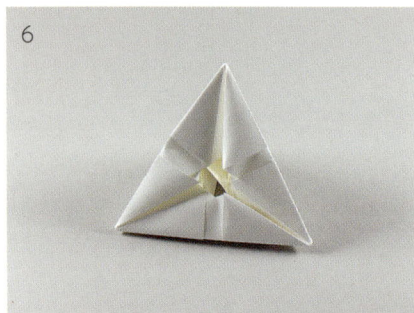

7

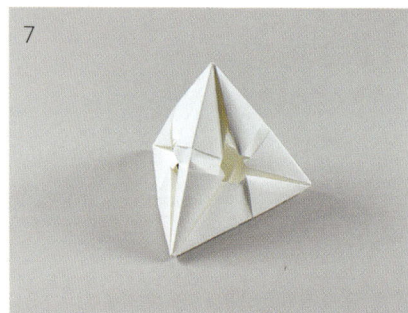

8

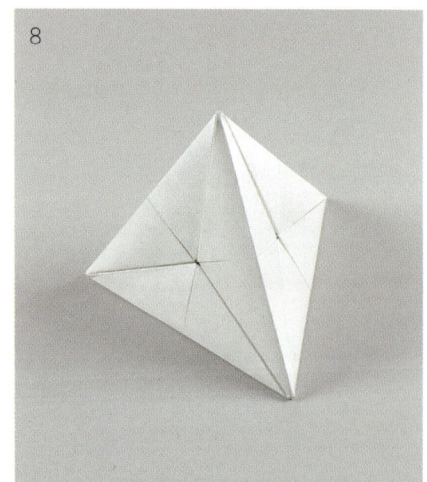

Tetrahedron

From Base to Module

You will need a total of 8 triangles of different sizes (4 × smaller horse and 4 × bigger rider). Note the sizes listed in the table on p. 32. Fold the mountain and valley folds according to the illustrations on the right (fig. 1). Note here the additional mountain folds. Join horse and rider into modules (figs. 2–3). Then glue the valley folds individually into the back of the additional mountain folds (fig. 4).

From Module to Finished Model

Join one module on one side with another module, pushing the modules only a small way into each other. Add additional modules individually to form a total of four trinomial ring joints (fig. 5). Only at the very end should you fully join all modules, going around the model several times and pushing them closer together little by little (figs. 6–8).

Difficulty: Demanding

Special challenges through additional folds, wraps, and gluing, as well as in the construction process; time required: ±1½ hours

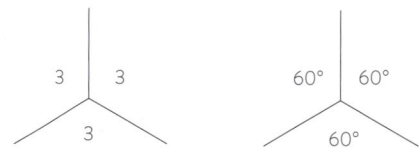

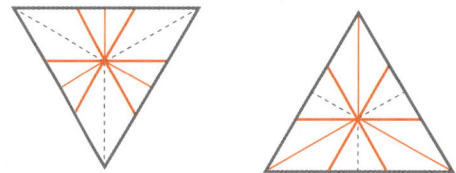

Note:

In the tetrahedron, only 3 × 60° are needed (i.e., out of 360°, 180° are superfluous here). This means 60° on each side have to be folded inwardly. The additional mountain folds we already applied in the construction of the cube are again used here. Additionally, the folded areas have to be wrapped inwardly and glued into the inside of the model.

4 x   4 x

F-models
Centaurs—Combined Horse/Rider Function

Some polyhedra require special angle sums between the edges that can be achieved only by putting together and folding two identical square, pentagonal, or hexagonal bases with functional cuts/slits according to the step-by-step instructions. Each of the two bases in the joint module then has areas that are on top as well as areas that are below. Due to the double horse/rider function, these sheets—after the horse/man beings in Greek mythology—are called "centaurs."

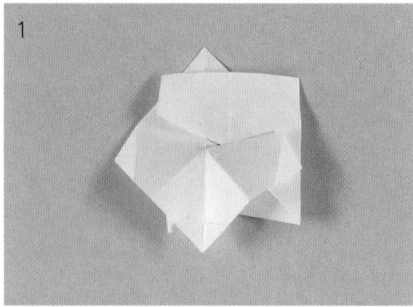

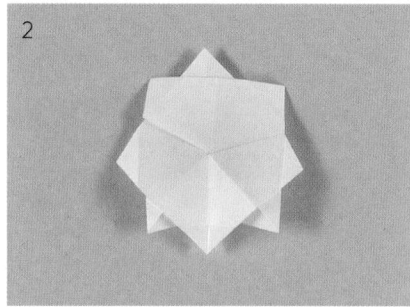

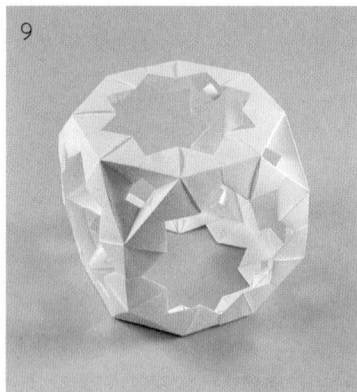

Truncated Cube

From Bases to Module

You will need a total of 48 squares of the same size (24 centaurs left and 24 centaurs right) that all are slit from one side (see illustration on the right and fig. 1). Fold mountain and valley folds according to the illustrations on the right and cut a slit, as illustrated, on the left (for centaur left) or on the right (for centaur right). Join one centaur left and one centaur right each to form a module (figs. 1 and 2). Then fold over the upper, protruding corners to complete the module (figs. 3 and 4).

From Module to Finished Model

Form eight trinomial ring joints (each of these forms an eighth of the finished model (figs. 5 and 6). Then put these together step by step (figs. 7–9).

Difficulty: Demanding

Special challenges due to detailed centaur modules, slits, and glue, as well as in construction; time required: ±2½ hours

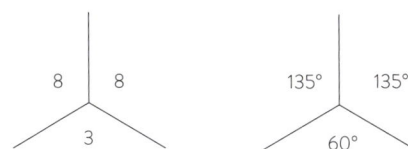

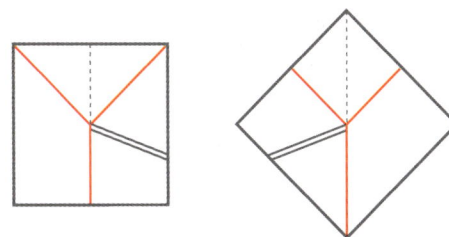

Note:
Any joints that are not stable can be secured with a small amount of glue.

24 x   24 x

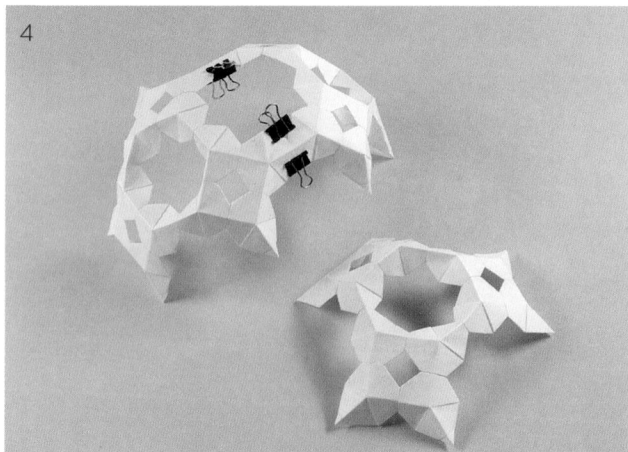

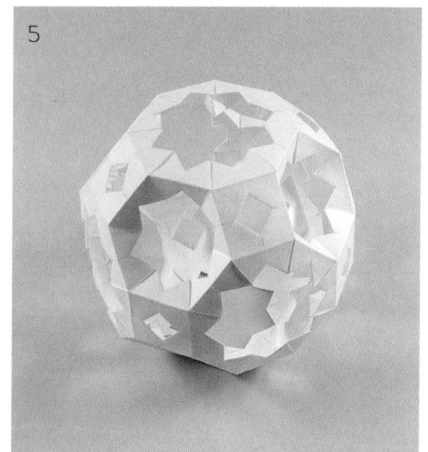

Truncated Cubo-octahedron

From Bases to Modules

You will need a total of 96 squares of the same size (2 × 24 centaurs left and 2 × 24 centaurs right) that all are slit from one side (see illustrations on the right). Fold mountain and valley folds according to the illustration on the right, then cut a slit on the left (for centaurs left) or on the right (for centaurs right), as illustrated (see figs. 1 and 2; see also preceding model). Then join one centaur left with one centaur right each to form a module.

Please note: you will need two mirror-image modules each (4 modules in total) that are joined alternatingly into quadrinomial ring joints (each of these forms a twelfth of the model).

From Module to Finished Model

Join the ring joints one after the other. Use glue and allow the joints to dry well (use paper clips to secure the construction; fig. 3). This results alternatingly in new six- and eightfold ring joints. In the same manner, complete the truncated cubo-octahedron as illustrated (figs. 4 and 5).

Difficulty: Demanding

Special challenges in detailed centaur modules, slits, and glue, as well as in construction; time required: ±4½ hours

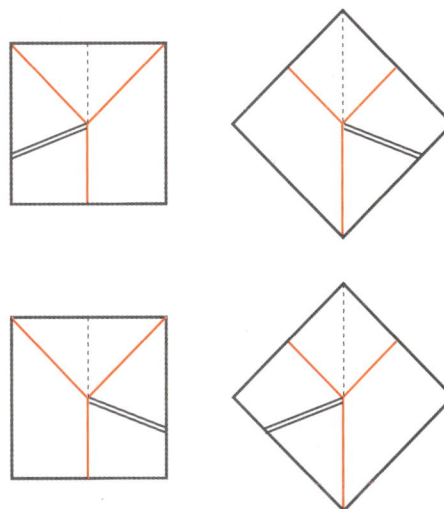

Note:

For the sixfold ring joints, no valley folds are created beforehand. In constructing the final model, "round" valley folds result almost by necessity when the superfluous material is pushed inside.

24 x   24 x   24 x   24 x

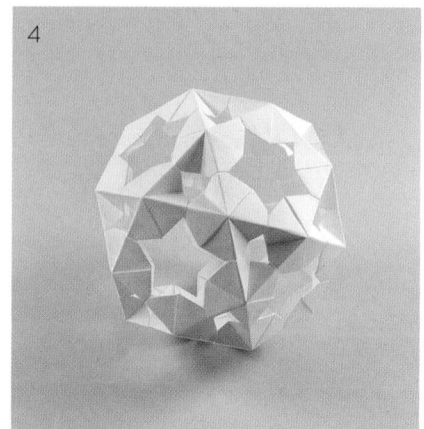

5	3		108°	60°
3	5		60°	108°

Icosidodecahedron

From Bases to Module

You will need a total of 60 pentagons of the same size (30 centaurs left and 30 centaurs right) that all are slit from one side (see illustrations on the right). Fold mountain and valley folds according to the illustrations on the right and cut a slit on the left (for centaur left) or right (for centaur right), as illustrated. Then join a centaur left and a centaur right each to form a module (fig. 1, *left*).

From Module to Finished Model

Form a first trinomial ring joint (fig. 1, *right*). Then join additional modules step by step. This results in two trinomial and two fivefold ring joints at each vertex of the model (fig. 2). In the same manner, complete the icosidodecahedron as illustrated (figs. 3 and 4).

Difficulty: Demanding

Special challenges through detailed centaur modules, cuts, and glue, as well as in construction; time required: ±3¼ hours

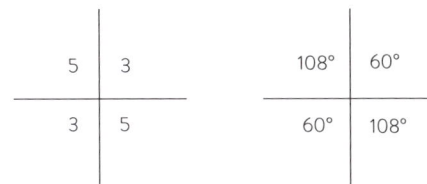

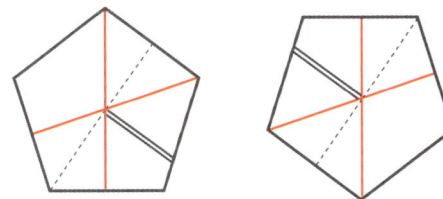

Note:
Any joints that are unstable can be secured using a small amount of glue.

30 x   30 x

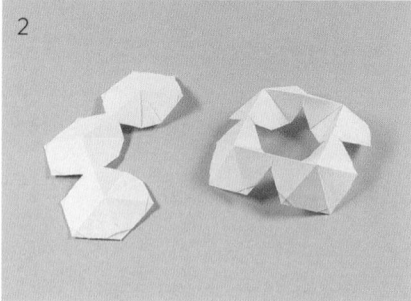

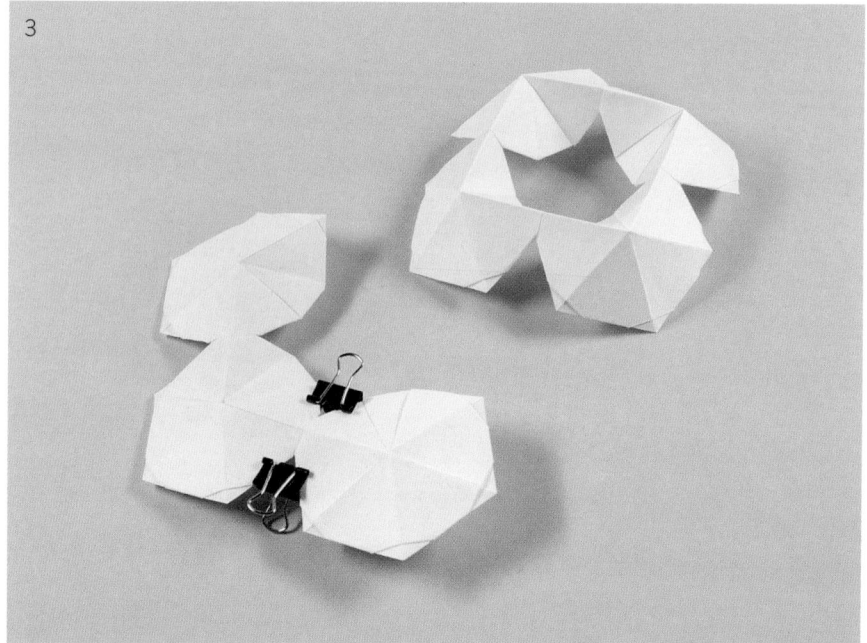

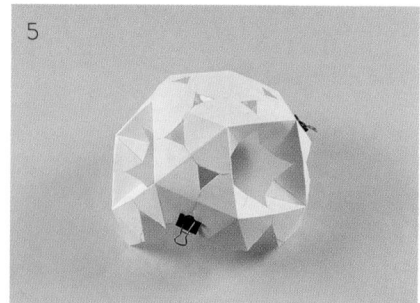

Cubus Simus
(Snub Cube; Chiral)

From Bases to Module
You will need a total of 48 hexagons of the same size (24 centaurs left and 24 centaurs right), each of which is slit from one side (see illustrations on the right). Fold the mountain and valley folds according to the illustrations on the right and cut a slit as illustrated on the left (for centaur left) or right (for centaur right). Then join one centaur left and one centaur right to form a module (fig. 1, *left*; *below*).

From Module to Finished Model
Form a first quadrinomial ring joint with the first four modules (this constitutes a sixth of the finished model; fig. 1, right through to fig. 3). Then add additional quadrinomial ring joints step by step (figs. 4 and 5). Two chiral models are possible of the cubus simus (fig. 6; see also models D14 and D15, page 70 and following).

Difficulty: Demanding
Special challenges due to cuts/slits, detailed centaur modules, and gluing, and in putting together; time required: ±4 hours per model

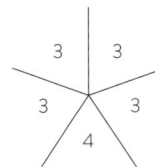

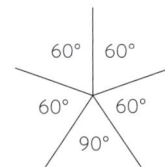

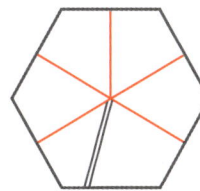

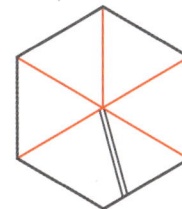

24 x   24 x

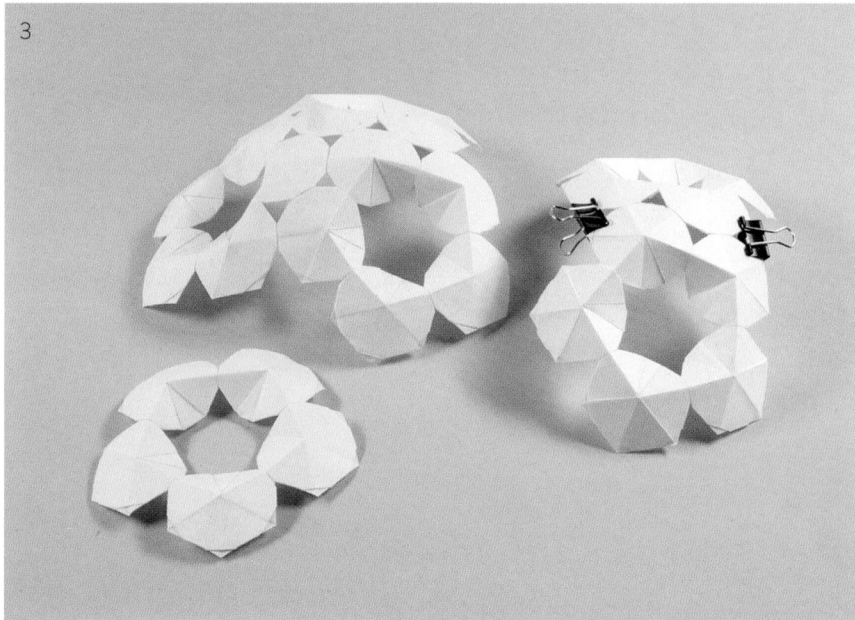

Dodecahedron Simum (Chiral)

From Base to Model
You will need a total of 120 hexagons in the same size (60 centaurs left and 60 centaurs right) that are slit from one side, respectively (see illustrations on the right). Fold the mountain and valley folds according to the illustrations on the right and cut a slit as illustrated on the left (for centaurs left) or right (for centaurs right). Then join a centaur left and a centaur right each to form a module (fig. 1, *left*).

From Module to Finished Model
Create a first pentamerous ring joint with the first five modules (this forms a twelfth of the model; fig. 1, *right*). Then put together additional fivefold ring joints step by step (figs. 2 and 3). Two chiral variants are possible of the dodecahedron simum (fig. 4; see also models D14, D15, and F 21, pp. 70, 72, and 88).

Difficulty: Very Difficult
Special challenges through detailed centaur modules, slits, and gluing, as well as in construction; time required: ±6 hours per model

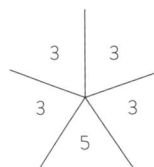

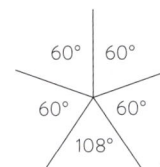

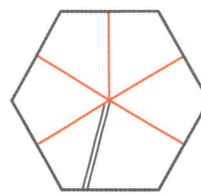

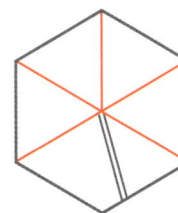

Note:
No valley folds are made for the pentamerous ring joints. In joining the model together, "round" valley folds appear almost by themselves when the excess material is pushed inward. Use a drop of glue when joining adjacent modules, and arrest the modules with clips until the glue has dried.

60 x   60 x

G-models
Complex Centaurs

To increase the challenge, this chapter presents models where a slit is cut in the base twice or three times. The construction of these centaurs is rather elaborate. Some of the models have a particularly large number of parts; consequently, they require not only a talent in working with one's hands, but also appreciably more patience.
You should attempt these forms only when you have successfully realized the models in chapter F.

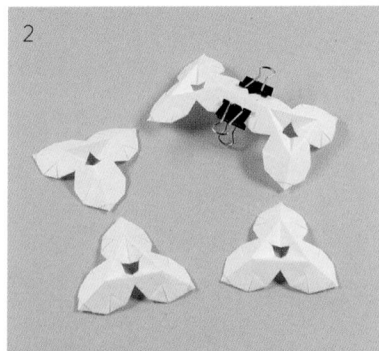

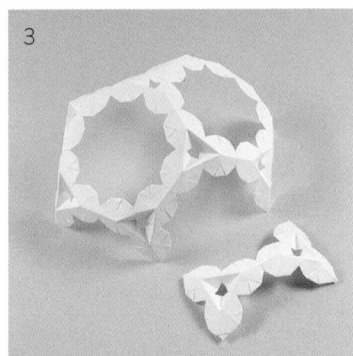

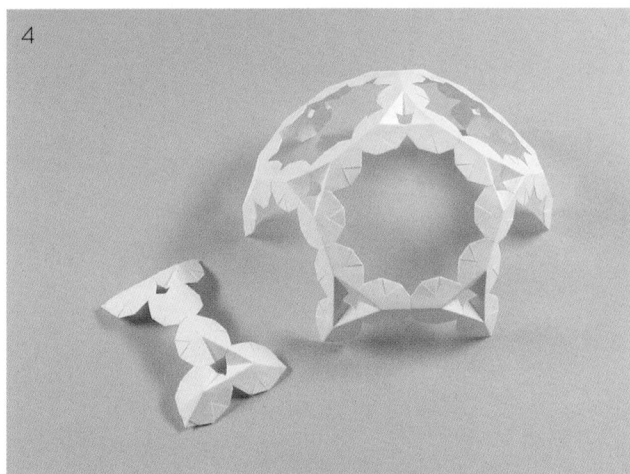

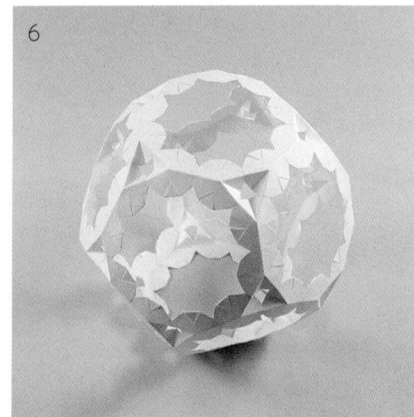

Truncated Dodecahedron

From Bases to Modules

You will need a total of 120 pentagonal bases in the same size (60 centaurs left and 60 centaurs right), each of which is slit from two sides (see illustrations on the right). Fold mountain and valley folds according to the illustrations on the right and cut a slit, as illustrated, on the left (for centaur left) or right (for centaur right). Then join a centaur right and a centaur left each to form a module (fig. 1).

From Module to Finished Model

Taking three modules, form a trinomial ring joint (this constitutes a twentieth of the finished model). Then join further trinomial modules. This results in new tenfold ring joints (figs. 2 and 3). Continue in the same manner (figs. 4–6) and complete the truncated dodecahedron. At each vertex of the finished model, two tenfold and a trinomial module meet.

Difficulty: Very Difficult

Particular challenges due to slits, detailed and complex centaur modules, and gluing, as well as in constructing; time required: ±6 hours

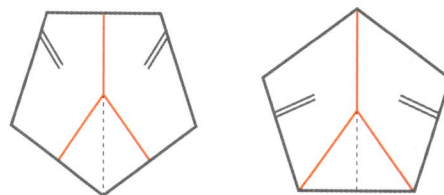

Note:

For the G-models, a single slit is not enough to distribute all horse-and-rider functions across the two bases. This makes one or two additional slits necessary. Any instable joints can be secured using a small amount of glue.

60 x   60 x

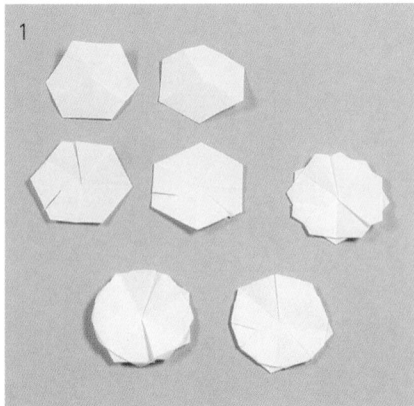

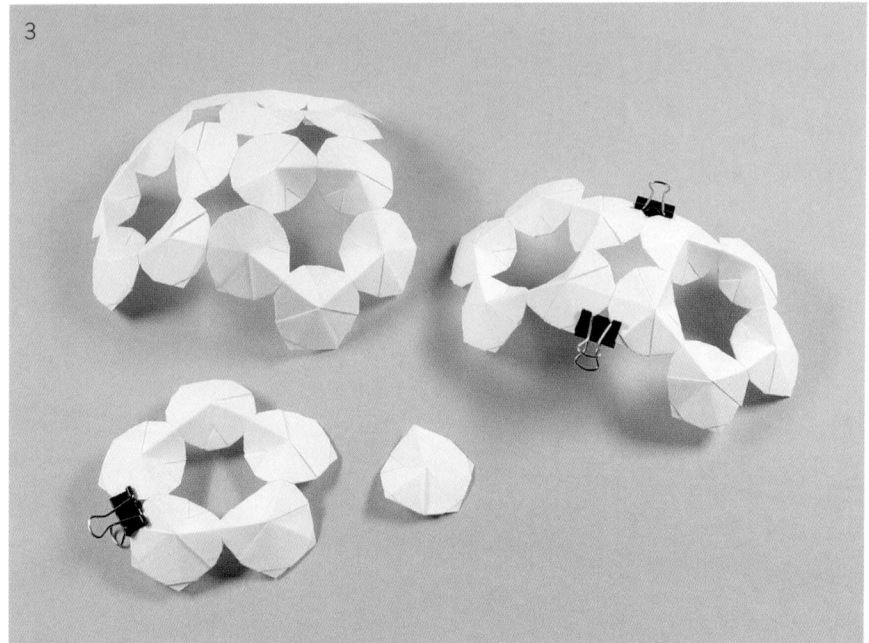

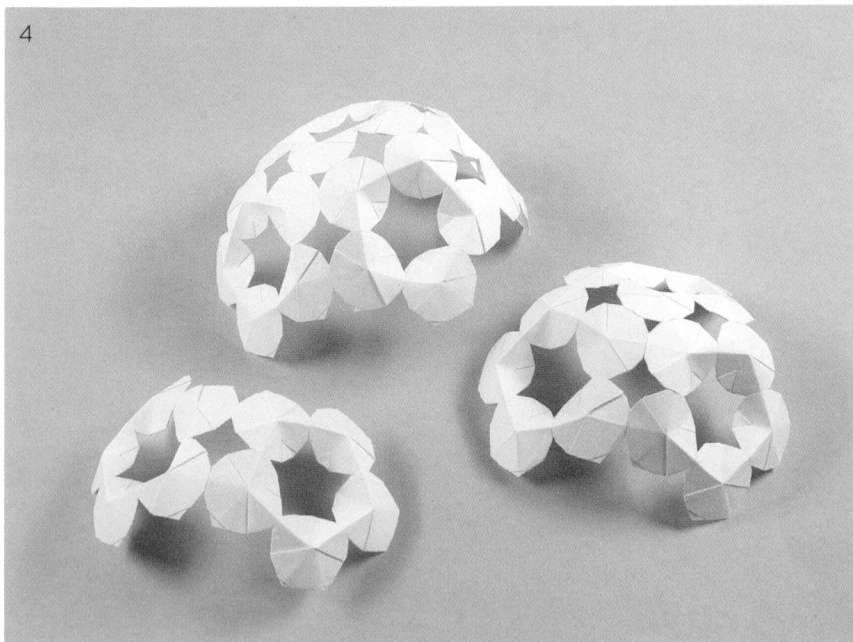

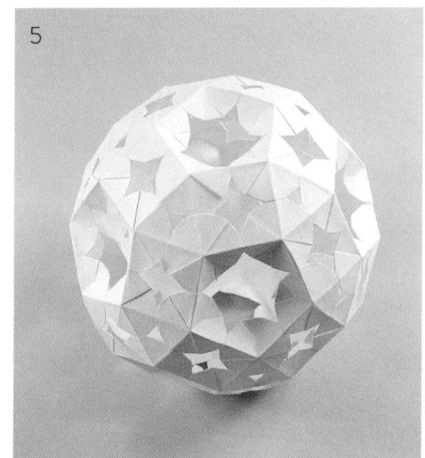

5	4	108°	90°
4	3	90°	60°

Rhombic Icosidodecahedron

From Bases to Module

You will need a total of 120 hexagons of the same size (60 centaurs left and 60 centaurs right), each of which is slit from two sides (see illustrations on the right). Fold mountain and valley folds according to the illustrations on the right and cut slits as illustrated on the left (for centaur left) and right (for centaur right). Join a centaur left and a centaur right each to form a module (fig. 1).

From Module to Finished Model

Form a trinomial ring joint from five modules each (fig. 2). Then join the pentamerous modules as illustrated. New tri- and quadrinomial ring joints are the result. Continue in the same manner (figs. 3 and 4) and complete the rhombic icosidodecahedron. At each vertex of the finished model, there is a total of two fourfold as well as a three- and a fivefold module.

Difficulty: Very Difficult

Particular challenges due to slits, detailed and complex centaur modules, and glue, as well as in construction; time required: ±8 hours

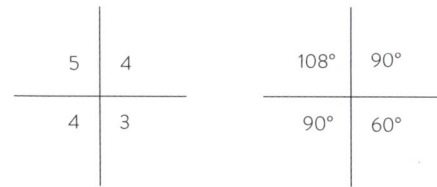

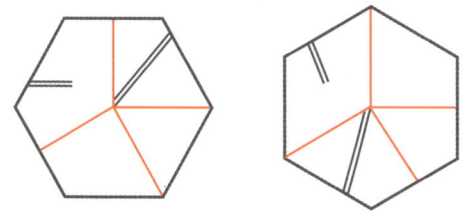

Note:

For the pentamerous ring joints, initially no valley folds are folded. When putting the model together, "round" valley folds result almost by themselves when the excess material is folded inward. Instable joints can be secured using a small amount of glue.

60 x   60 x

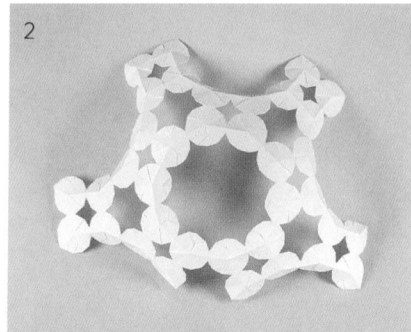

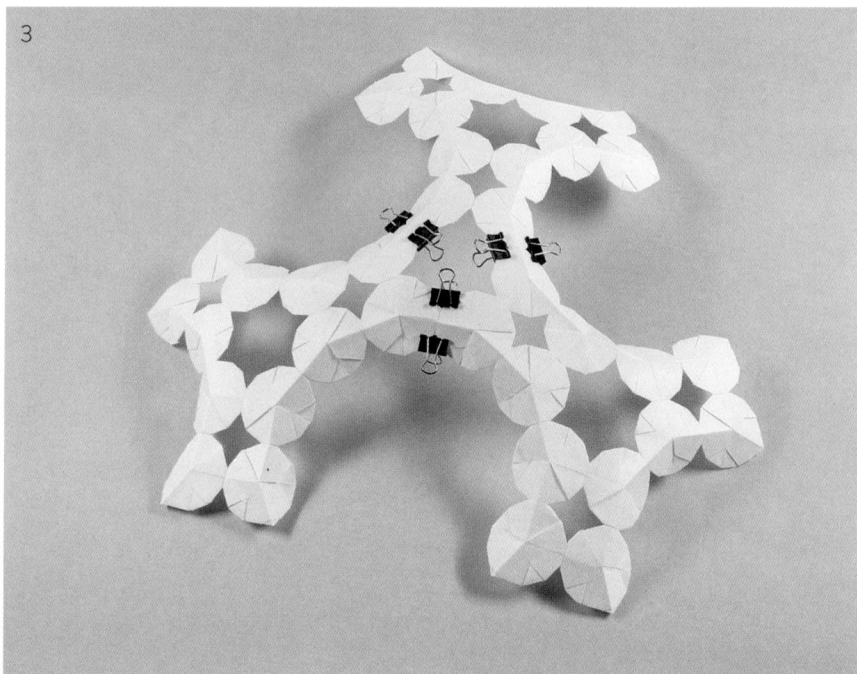

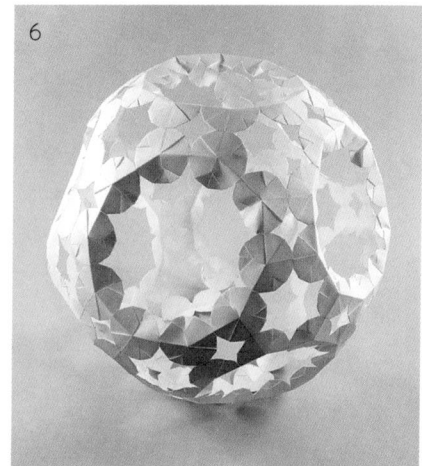

Truncated Icosidodecahedron

From Bases to Module

You will need a total of 240 hexagons of the same size (2 × 60 centaurs left and 2 × 60 centaurs right), with three slits each. Fold mountain and valley folds according to the illustrations on the right and cut a slit as illustrated on the left (for centaur left) or on the right (for centaur right). Join a centaur left and a centaur right each to form a module. Please note: you will need two mirror-image modules joined with each other to form 30 quadrinomial ring joints (fig. 1, *left*).

From Module to Finished Model

Then join the quadrinomial ring joints with each other as illustrated (fig. 1, *right*). New six- and tenfold ring joints are the result (fig. 2). Continue in the same manner (figs. 3 and 4). At each vertex of the finished truncated icosidodecahedron (fig. 5) a quadrinomial, a sixfold, and a tenfold module meet.

Difficulty: Most Challenging

Special challenges due to slits, complex and detailed centaur modules, and glue, as well as in construction. Building this model at a larger scale is marginally less difficult. Time required: ±20 hours

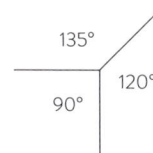

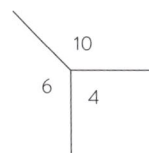

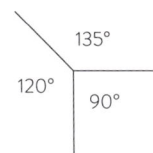

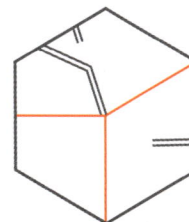

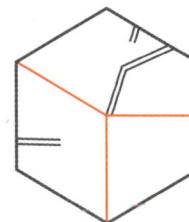

Note:

Do not create valley folds in tenfold ring joints. In the final construction, "round" valley folds occur somewhat by themselves (compare with preceding models). Any instable joints can be secured using a small amount of glue, as well as additionally with supports (compare with the following models).

60 x   60 x   60 x   60 x

H-models
with Supports

Modules with six, eight, or ten vertices cannot be joined as easily as those with fewer vertices. Special supports made from overhead projector foil are an effective help in this case. These supports are pushed between horse and rider of two neighboring modules, linking the modules and lending stability to the finished model.

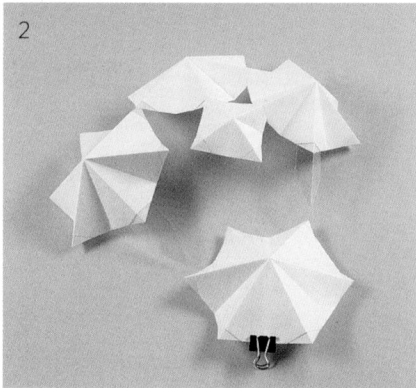

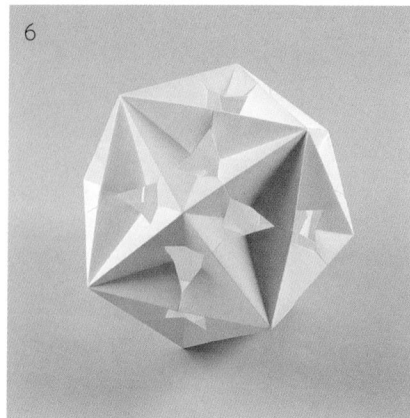

3	3
3	3

83°	83°
83°	83°

3	3
3	3
3	3

48°	48°
48°	48°
48°	48°

Tetrakis Hexahedron

From Bases to Modules

You will need a total of 12 squares of the same size (6 × horse and 6 × rider) as well as 16 hexagons in different sizes (8 × smaller horse and 8 × larger rider). Additionally you will need 12 supports. Fold mountain and valley folds according to the illustrations on the right. Then join horse and rider into modules (fig. 1).

From Modules to Finished Models

Add supports at every second vertex in one of the sixfold modules and join another sixfold module. You may have to use glue. Join quadrinomial modules at each of the other vertices (figs. 2–3). Continue in the same manner (figs. 4–6) and complete the model. Note the following principle: join quadrinomial modules exclusively with sixfold modules. The sixfold modules have alternating joints with quadrinomial and sixfold modules.

Difficulty: Challenging

Special challenges due to complex modules, supports, and glue; time required: ±2 hours

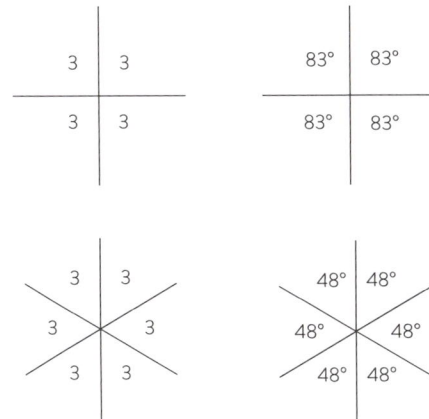

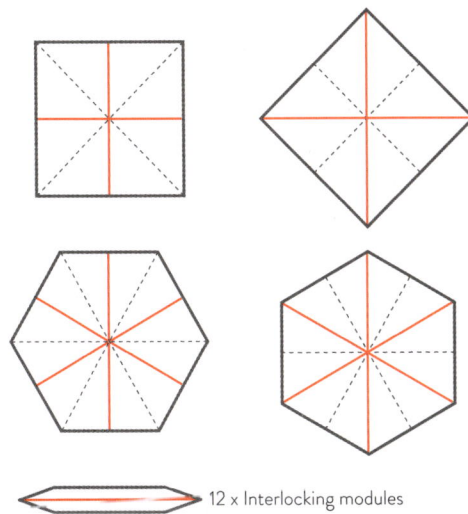

12 x Interlocking modules

Note:

A support that joins two modules inside the model is a good solution and increases the model's stability. Colorless and transparent overhead projector foil is useful here. It is used only to stabilize the joints between two hexagons, octagons, or decagons.

6 x	6 x	8 x	8 x

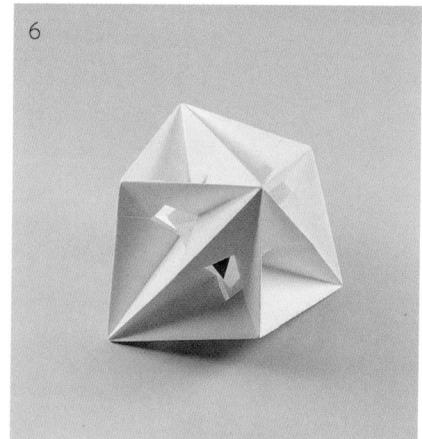

Triakis Tetrahedron

From Bases to Modules

You will need a total of 8 triangles in different sizes (4 ×
bigger horse and 4 × smaller rider) as well as 8 hexagons of
the same size (4 × horse and 4 × rider). Additionally you will
need 12 supports. Fold mountain and valley folds according
to the illustrations on the right. Then join horse and rider
into modules.

From Modules to Finished Model

Take one of the sixfold modules and put a support at every
second vertex. Then join a second sixfold module, gluing it
in place if necessary. Add trinomial modules at the remaining vertices (figs. 2–3). Continue in the same manner
(figs. 4–6) and finish the model. Note the following principle: join trinomial modules exclusively with sixfold modules.
Each sixfold module is continued with alternating three-
and sixfold modules.

Difficulty: Challenging

Special challenges due to complex modules, supports, and
glue; time required: ±2 hours

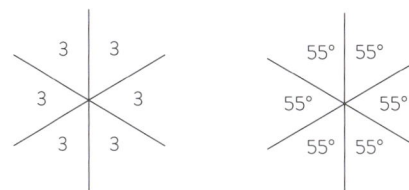

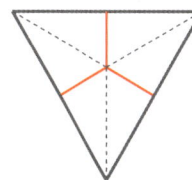

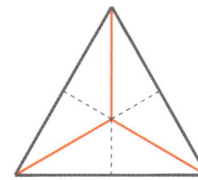

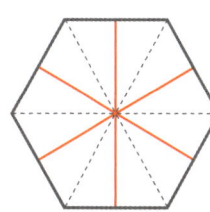

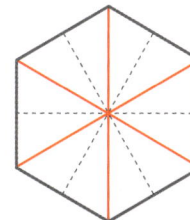

12 x Interlocking modules

Note:
In this model, supports are needed only
between two sixfold modules.

4 x   4 x   4 x   4 x

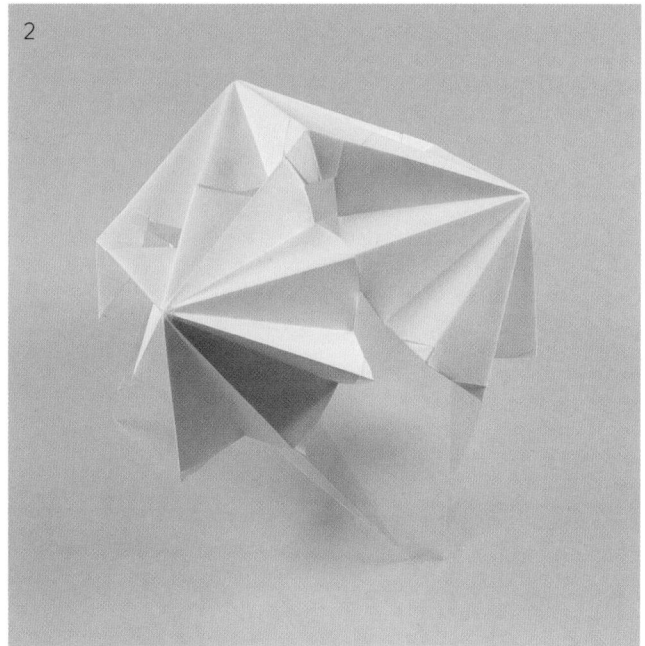

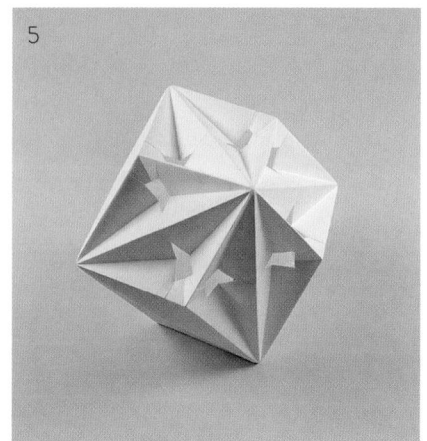

Triakis Hexahedron (3/8)

From Bases to Modules

You will need a total of 16 triangles in different sizes (8 × larger horse and 8 × smaller rider) as well as 12 octagons of the same size (6 × horse and 6 × rider). Additionally you will need 12 supports. Fold mountain and valley folds according to the illustrations on the right. Then join horse and rider into modules (fig. 1, *left*).

From Modules to Finished Model

Take one of the eightfold modules and put a support at every second vertex, then add another eightfold module, gluing it in place if necessary (fig. 1, *right*; fig. 2). Add threefold modules at the remaining vertices (figs. 2–3). Continue in the same manner.

Note the following principle: join trinomial modules only with eightfold modules. Each eightfold module has trinomial and eightfold modules alternately attached.

Difficulty: Challenging

Special challenges due to complex modules, supports, and glue; time required: ±3 hours

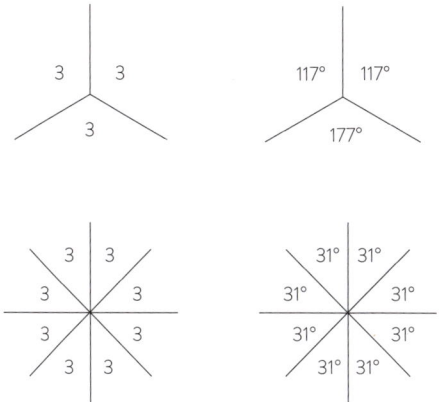

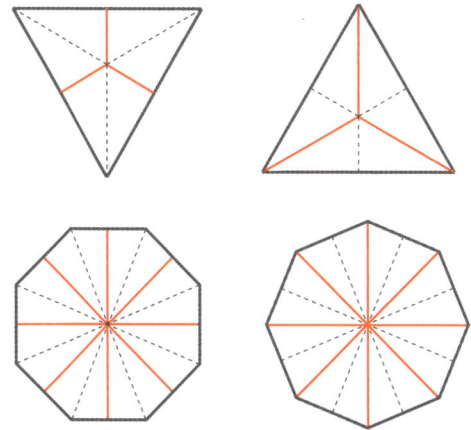

12 x Interlocking modules

8 x   8 x   6 x   6 x

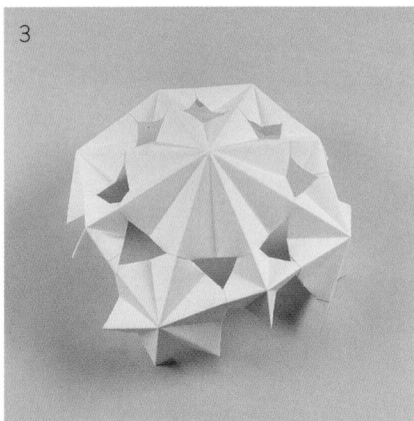

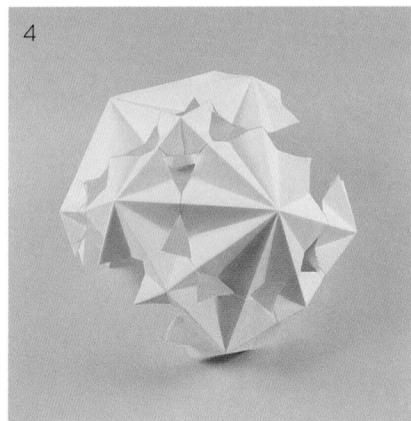

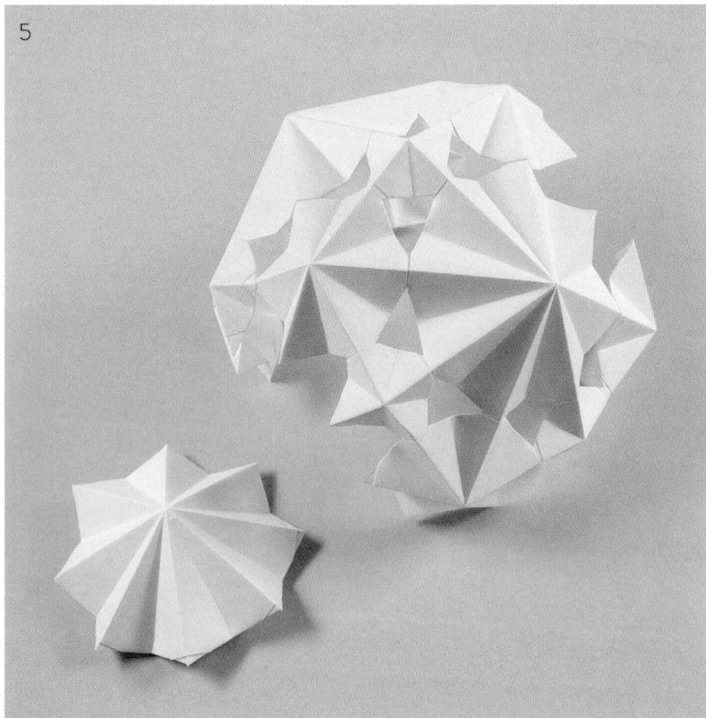

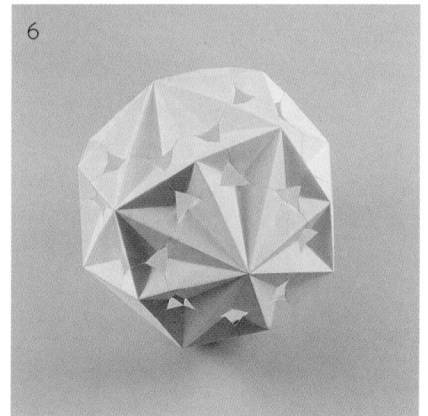

Triakis Hexahedron (4/6/8)

From Bases to Modules

You will need a total of 24 squares (12 × horse and 12 × rider), 16 hexagons (8 × horse and 8 × rider), and 12 octagons (6 × horse and 6 × rider). Additionally you will need 24 supports. Fold mountain and valley folds according to the illustrations on the right. Then join horse and rider to form modules (fig. 1, *left*).

From Modules to Finished Model

Take one of the eightfold modules and put a support into every second vertex, then add a second sixfold module, gluing it in place if necessary. Add quadrinomial modules at the remaining vertices (fig. 1, *center*). Continue in the same manner (figs. 2–6) and complete the model.

Note the following principle: join quadrinomial modules alternatingly with six and eightfold modules, sixfold modules alternatingly with quadrinomial and eightfold modules, and eightfold modules alternatingly with quadrinomial and sixfold modules. The supports are put between the eight- and the sixfold modules.

Difficulty: Challenging

Special challenges due to complex modules, supports, and glue; time required: ±4½ hours

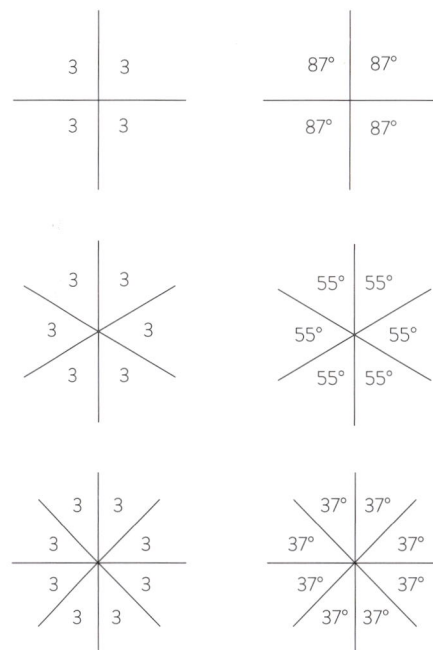

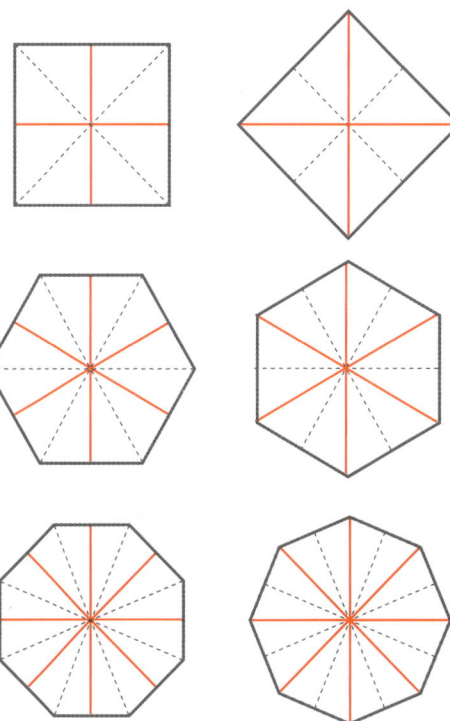

3	3		87°	87°
3	3		87°	87°

24 x Interlocking modules

12 x	12 x	8 x	8 x	6 x	6 x

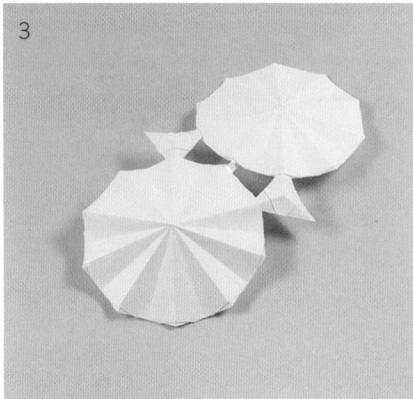

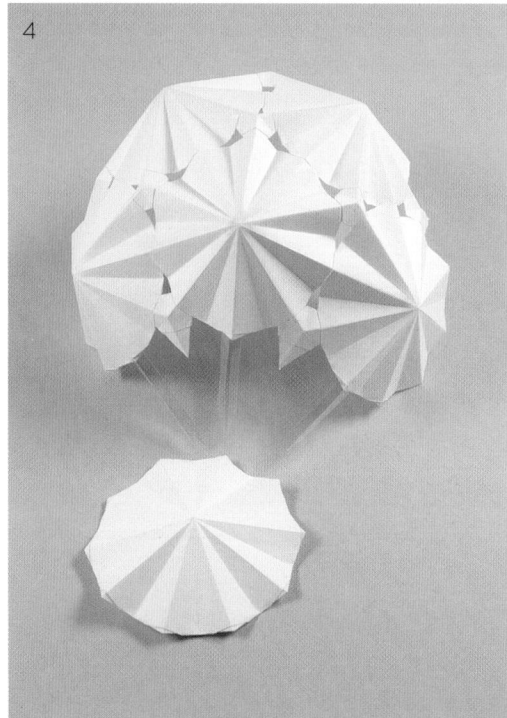

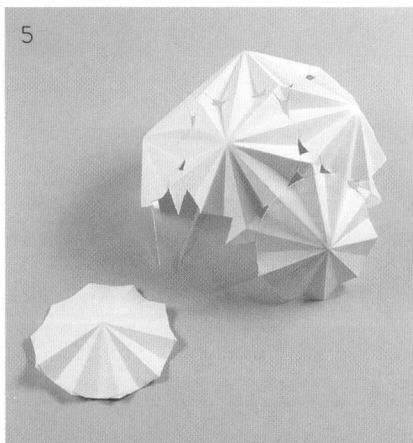

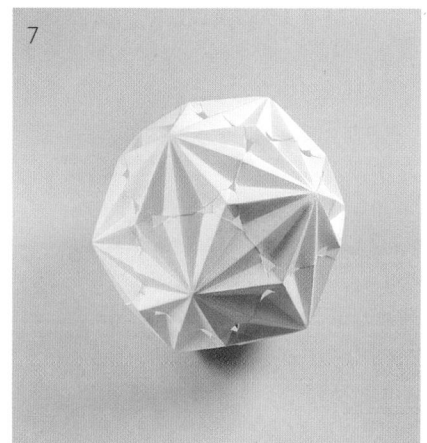

Pentakis Dodecahedron (3/10)

From Bases to Modules

You will need a total of 40 triangles in two sizes (20 × larger horse and 20 × smaller rider), as well as 24 decagons of the same size (12 × horse and 12 × rider). Additionally you will need 30 supports. Fold mountain and valley folds according to the illustrations on the right. Then join horse and rider to form modules (fig. 1, *left*). Put a dab of glue onto the folded-over tips of the riders in the tenfold modules and allow to dry, using clips before proceeding with the construction of the model (figs. 1 and 2).

From Modules to Finished Model

Take one of the tenfold modules and put a support in every second vertex, then join another tenfold module, gluing it in place if necessary. Join trinomial modules to the other vertices. Fig. 3 shows two trinomial and tenfold modules each. Continue in the same manner (figs. 4–6) and complete the model (fig. 7). Note the following principle: join trinomial modules exclusively with tenfold modules. In contrast, trinomial and tenfold modules are added alternatingly to tenfold modules.

Difficulty: Very Difficult

Special challenges due to complex modules, supports, and glue; time required: ±10 hours

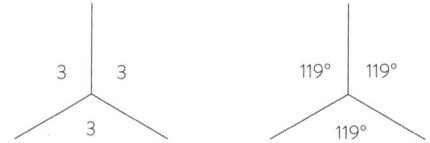

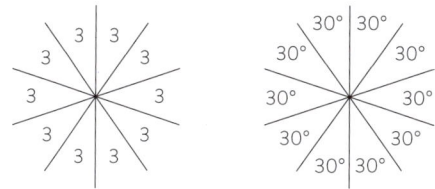

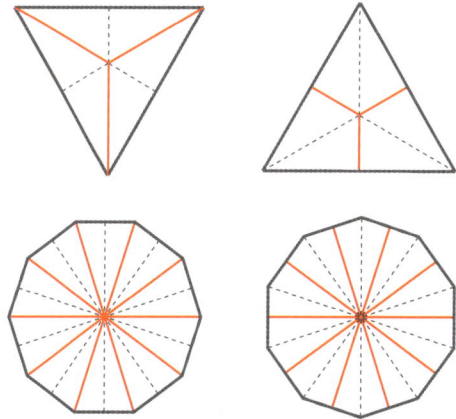

30 x Interlocking modules

20 x   20 x   12 x   12 x

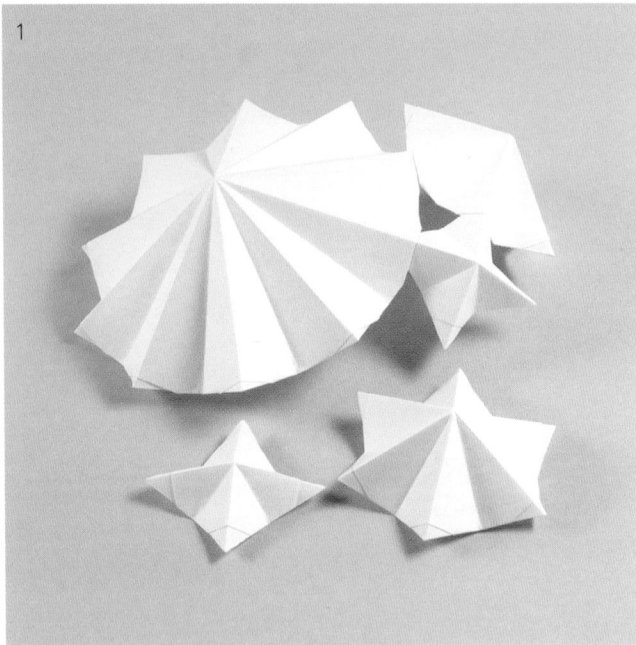

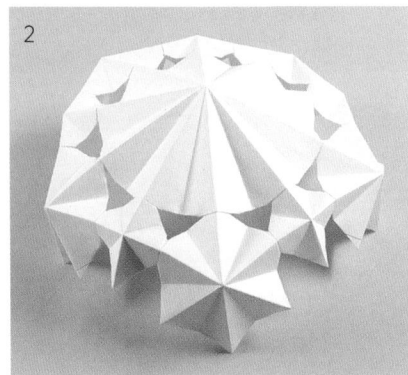

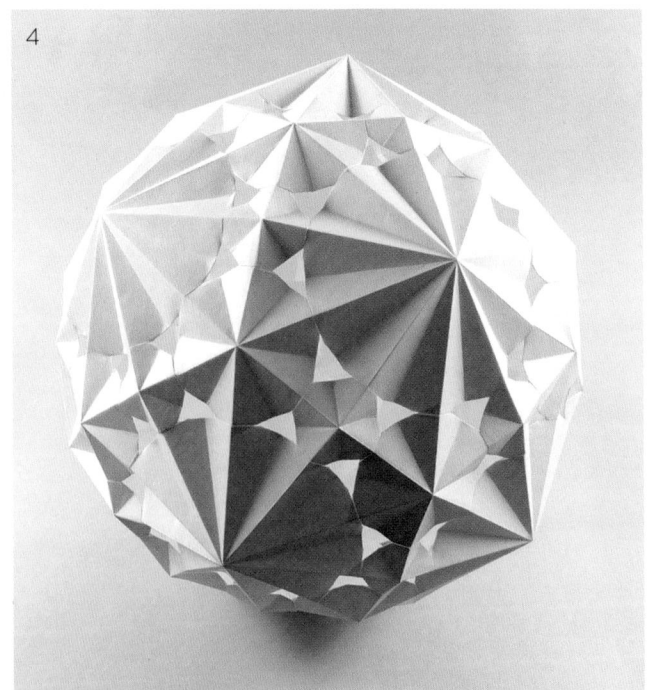

Pentakis Dodecahedron (4/6/10)

From Bases to Modules

You will need a total of 60 squares (30 × horse and 30 × rider), 40 hexagons (20 × horse and 20 × rider), and 24 decagons (12 × horse and 12 × rider). Additionally you will need 120 supports. Fold mountain and valley folds according to the illustrations on the right. Then join horse and rider to form modules (fig. 1).

From Modules to Finished Model

Take one of the tenfold modules and put a support into every second vertex, then add a sixfold module, gluing it in place if necessary. Add quadrinomial modules at the other vertices. Continue in the same manner (figs. 2–4) and complete the model.

Note the following principle: join quadrinomial modules alternatingly with six- and tenfold modules, sixfold modules alternatingly with quadrinomial and tenfold modules, and tenfold modules alternatingly with quadrinomial and eight-fold modules.

Difficulty: Most Challenging

Particular challenges due to complex modules, supports, and glue; time required: ±18 hours

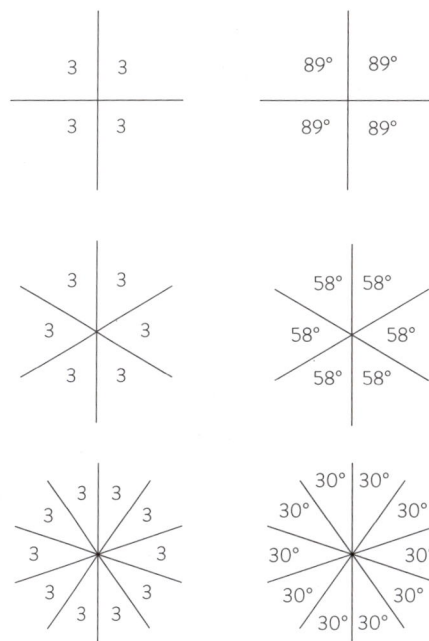

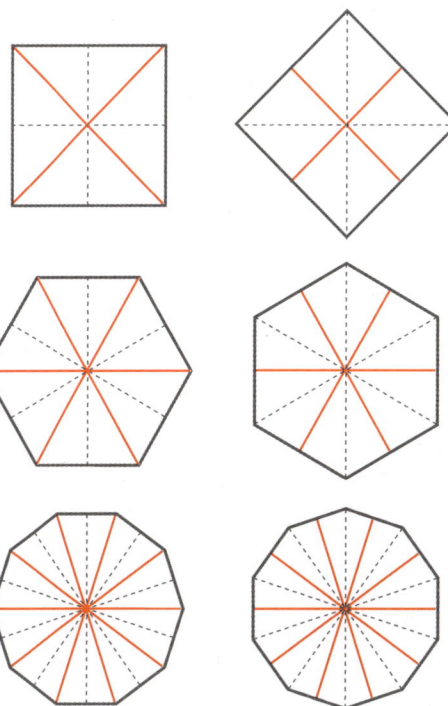

120 × Interlocking modules

| 30 x | 30 x | 20 x | 20 x | 12 x | 12 x |

I-models
New Chiral Forms

Analogous with models F21 (cubus simus with tri- and quadrinomial ring joints; see p. 88) and F22 (dodecahedron simum with tri- and pentanomial ring joints; see p. 90), you can construct two mirror-image (chiral) icosahedra with tri- and pentanomial ring joints using the same parts.

The same principle applies to model B04 (dodecahedron; see p. 46), which can be realized using two different triangles: using the triangles from B04 or the triangles of the E-models (see p. 74 and following)

Icosahedron
(Chiral)

From Bases to Modules

You will need a total of 24 hexagons (12 centaurs left and 12 centaurs right), each of which has been cut from one side (see illustrations on the right). Fold mountain and valley folds according to the illustrations on the right and cut a slit on the left (for centaur left) or on the right (for centaur right), as illustrated. Then join a centaur left and a centaur right each to form a module (fig. 1, *left*; see also models F21 and F22, p. 88 and following).

From Module to Finished Model

Take three modules and form a first trinomial ring joint (this forms a fourth of the finished model). Fig. 1 (*right*) shows the intermediate step. Form further trinomial ring joints and put these together step by step (figs. 2 and 3). Two chiral variants can be created of this model (fig. 4; see also models D14, D15, F22, and I22, pp. 70, 72, 88, 90, and 118).

Difficulty: Challenging

Particular challenges due to slits, complex centaur modules, and glue, as well as in constructing the final model; time required: ±2 hours per model

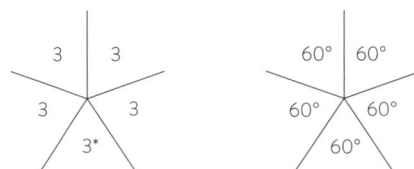

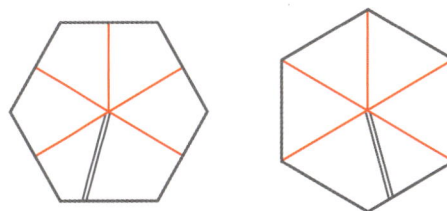

Note:

No valley folds are folded for the trinomial ring joints. In construction, "round" valley folds occur almost by themselves when the excess material is pushed inward.

12 x   12 x

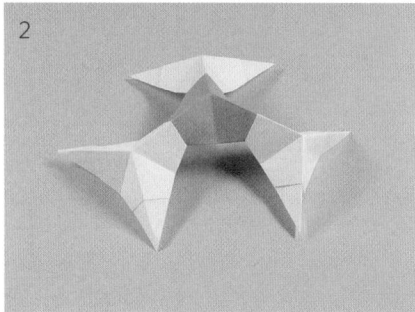

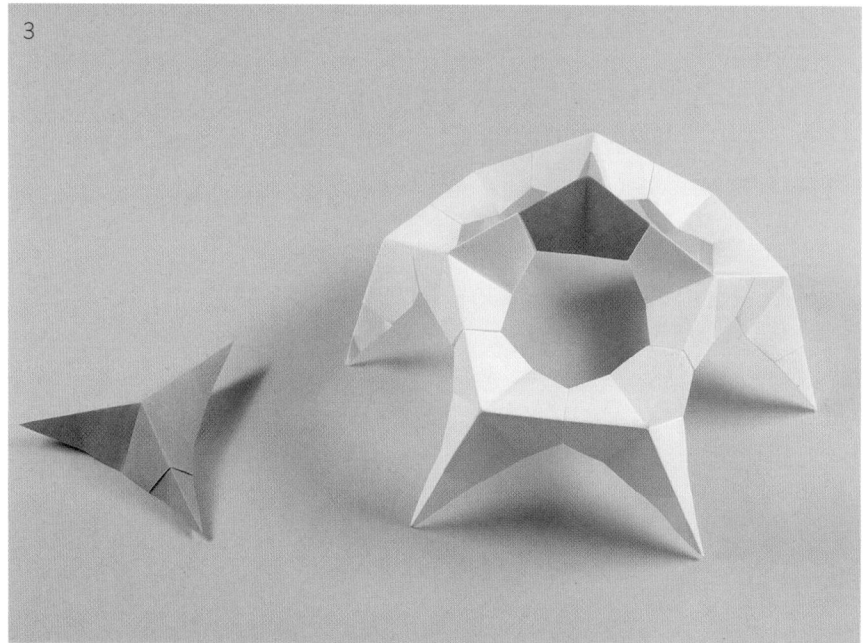

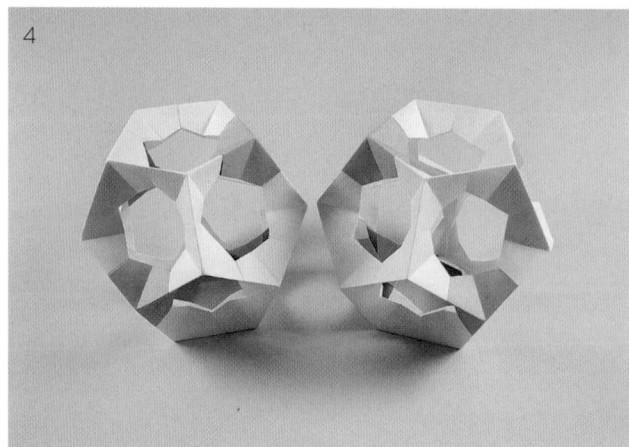

Dodecahedron
(Chiral)

From Base to Module

You will need a total of 32 light and 8 dark triangles of the
same size, consisting of 16 × horse and 16 × rider in a light color
as well as 4 × horse and 4 × rider in a dark color. Fold mountain
and valley folds according to the illustrations on the right and
join to form single-colored modules (fig. 1). Then join each dark
module exclusively with light modules (figs. 2–4).

From Module to Finished Model

Put all parts formed up to this point together in a way similar
to the procedure used for model B04 (see p. 46). Be careful
to distribute light and dark modules evenly across the model.
Two chiral variants can be built of this model (fig. 4; see also
models D14, D15, F21, F22, and I32, pp. 70, 72, 88, and 90,
as well as 116).

Difficulty: Medium

Simple cutting, some additional folds, simple construction;
time required: ±1 hour per model

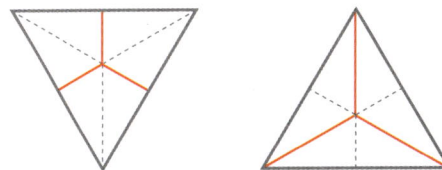

Extensions of Polyhedra Forms

The form diversity of the folded polyhedra presented in chapters A–I is limited to some thirty models. If we loosen the strict numerical specifications at their basis, it becomes possible to experiment freely while still using the same horse-and-rider technique. Only the simplest models that can be created from horse-and-rider modules of the same size are presented in the following chapter. The triangles, squares, and pentagons of the bases for all models are of the same size and can be freely combined.

Illustration of Base and Module

As before, you will find schematic illustrations of the bases of horse and rider on the right-hand side of the step-by-step instructions for each model. These show what type of base is required for the respective modules and how to fold the bases. Different types of modules are additionally realized in contrasting colors across the entire section "Extensions of Polyhedra Forms," as the pictures (figures) accompanying each model clearly show.

Nearly all extended forms are easy to construct. This is to stimulate you to continue experimenting on your own. This is also the reason why I have kept the step-by-step instructions simple. If you would like to go beyond these simple, uniformly sized forms, you can easily find suggestions in the instructions for the preceding polyhedra. This should enable you to broaden the spectrum of forms considered here using horse-and-rider bases of different sizes and centaurs, or wrapped and enveloped bases.

All shapes in the following chapter can be built easily and do not present any particular challenges. They do not exceed the difficulty of models A, B, and C (see p. 36 and following). Allow yourself to be motivated to try many more-playful experiments! Those who do not feel challenged enough will find further possibilities of combinations from chapters D onward (see p. 64 and following).

Uniform Color Coding

To facilitate orientation, a uniform color code was used across the following models. Square modules were done in dark blue if all parallel edges and diagonal folds were folded. If one valley fold is missing entirely, the module is shown in orange. A module with only one half-valley fold is dark yellow. A module with two half-valley folds is light green, and a module with three half-valley folds is dark red.

It is similar with bases from triangles, pentagons, and hexagons. The color coding should allow you to see at one glance how a module is to be folded or which bases it uses.

Uniform Designation of Models

4 = quadrinomial ring joint, square
4* = quadrinomial ring joint, quadrangular but not square
6' = sixfold ring joint with concave fold
6" = sixfold ring joint with two concave folds
3** = incomplete triangle (in torso)

J-models
Extensions of the Cubo-octahedron and the Rhombic Cubo-octahedron

All models introduced in this chapter are based on the polyhedra forms cubo-octahedron (model A02; see p. 40) and rhombic cubo-octahedron (model A03; see p. 42). In the following models, individual parts of these basic models were put in a different place or extended by changing parts slightly, or leaving out parts. The resulting model form usually very clearly shows its relationship with a particular polyhedron form from the preceding chapter.

To facilitate orientation, the relevant basic model is mentioned in the instructions or can be easily deduced from the name of the free form.

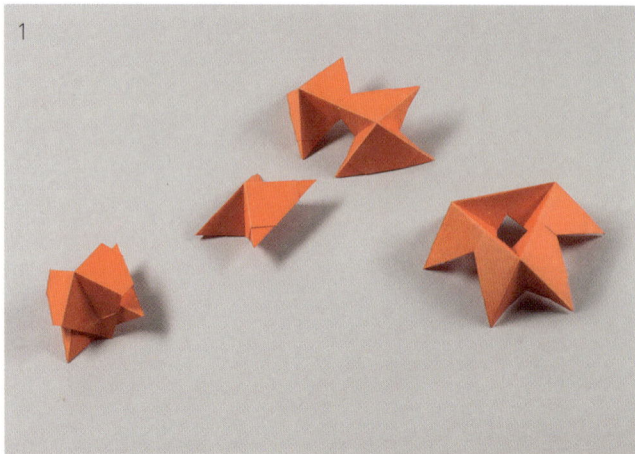

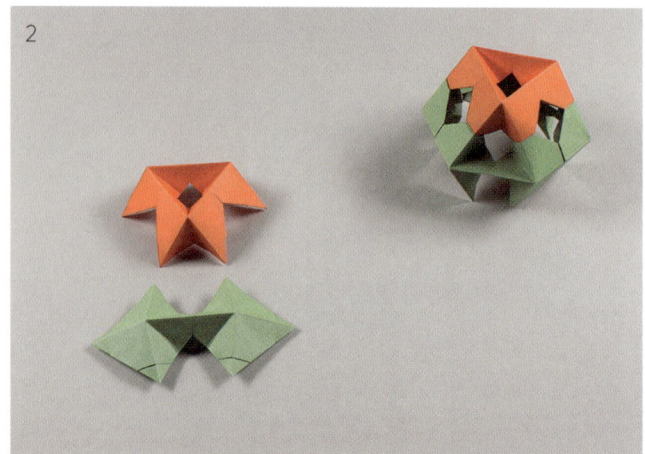

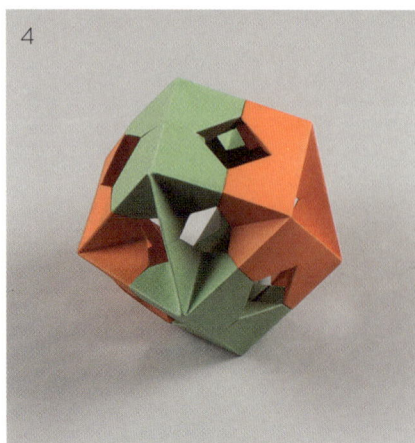

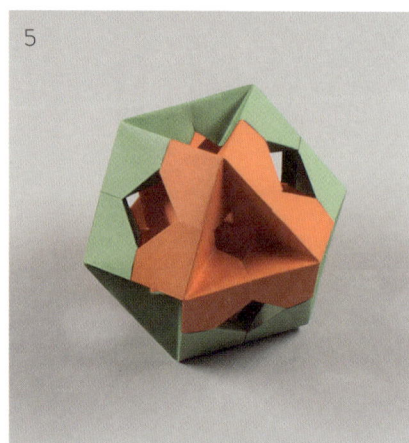

Twisted Cubo-octahedron

From Base to Module

You will need a total of 24 squares, consisting of 6 × horse and 6 × rider in orange as well as 6 × horse and 6 × rider in green; both bases have the same size. Fold mountain and valley folds according to the illustrations on the right. Please note: the green and orange bases have different mountain and valley folds.

Then join horse and rider into modules of the same color.

From Module to Finished Model

Join the first three orange modules to form a trinomial ring joint (fig. 2) and repeat this step. Then join the green modules—attached to orange modules—to form a green band (fig. 2). Finally, add a second orange ring joint to the green modules; this completes the model (fig. 3). The resulting form (figs. 4 and 5) corresponds to two halves of the cubo-octahedron (model A02; see p. 40) that are twisted diametrically.

Difficulty: Easy

Simple cutting, simple construction; time required: ±1 hour

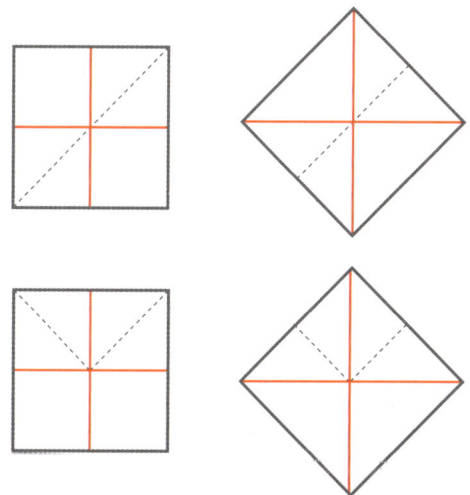

6 x   6 x   6 x   6 x

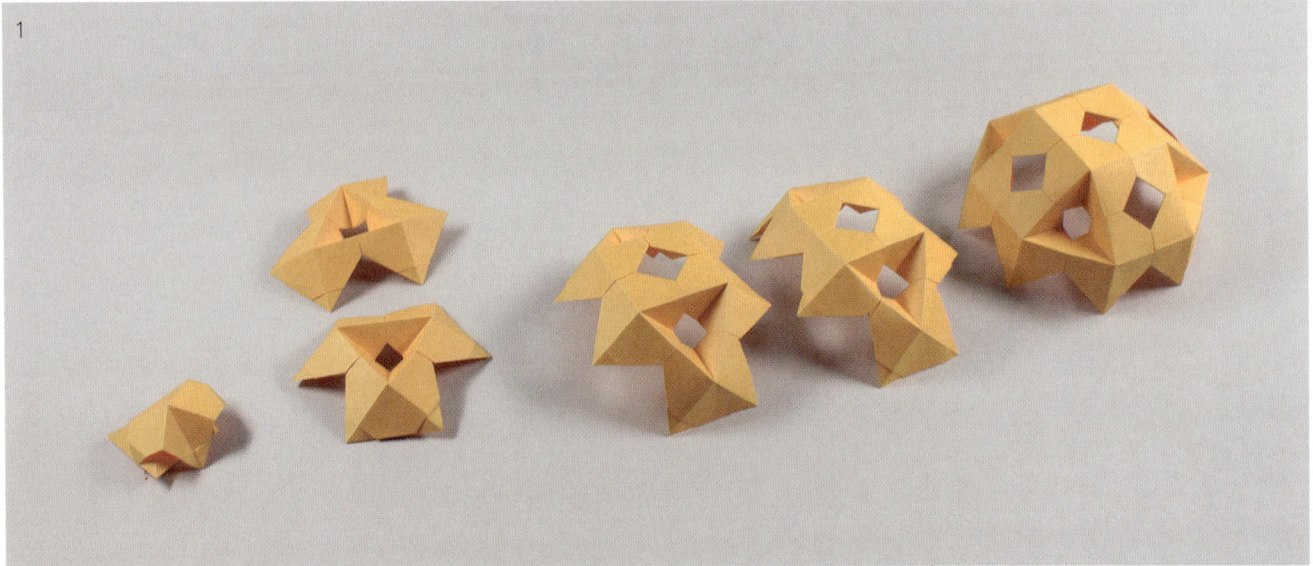

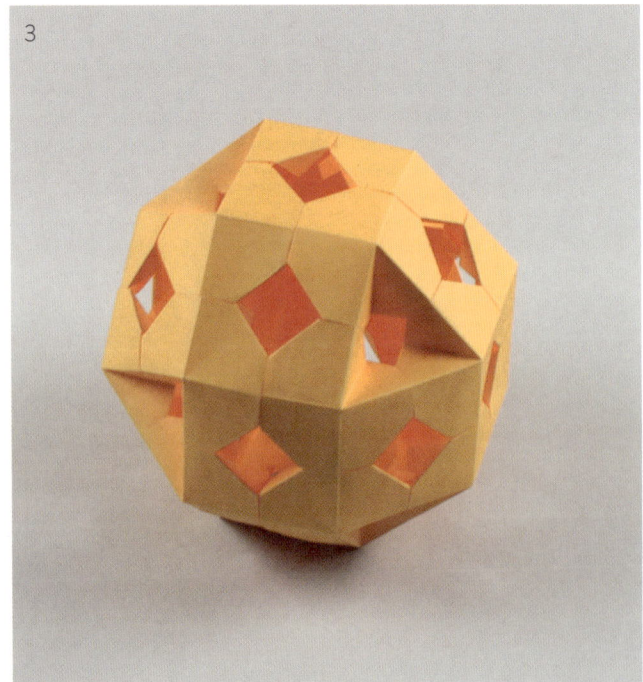

Twisted Rhombic Cubo-octahedron (3/4)

First Steps

You will need a total of 48 squares, consisting of 24 × horse and 24 × rider of the same size. Fold mountain and valley folds according to the illustrations on the right. Then join horse and rider to form modules and join these to form trinomial ring joints (fig. 1, *from left to right*).

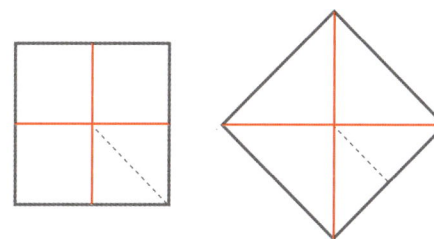

Constructing the Model

Join four trinomial ring joints to form the two halves of the polyhedron (fig. 1, *right*; fig. 2) and push these together. The resulting form (fig. 3) corresponds to two diametrically twisted halves of the rhombic cubo octahedron (model A03; see p. 42).

Difficulty: Easy

Simple cutting, simple construction; time required: ±75 minutes

24 x   24 x

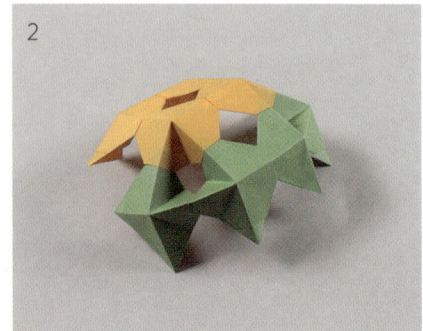

Rhombic Cubo-octahedron Cap

From Base to Module

You will need a total of 32 squares, consisting of 8 × horse and 8 × rider in yellow as well as 8 × horse and 8 × rider in green—both bases of the same size. Fold mountain and valley folds according to the illustrations on the right. Please note: the yellow and green bases have different mountain and valley folds. Then join horse and rider of the same color to form modules (fig. 1).

From Module to Finished Model

Join the first four yellow modules to form a quadrinomial ring joint (fig. 1, *right*), and repeat this step. Then join the green modules—attached to yellow modules—to form a green band (figs. 2 and 3). Finally, add a second yellow fourfold ring to the green modules. This completes the model.

The resulting model looks like two caps (or lids) of the rhombic cubo-octahedron (model A03; see p. 42) that have been put together. Fig. 4, *left*, shows the finished model next to the "twisted rhombic cubo-octahedron cap," as described in the following model (see p. 130).

Difficulty: Easy

Simple cutting, simple construction; time required: ±1 hour

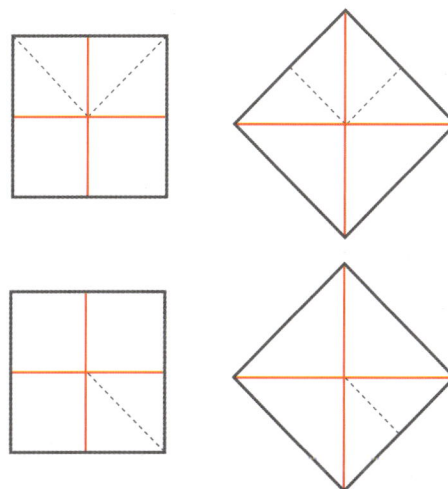

8 x   8 x   8 x   8 x

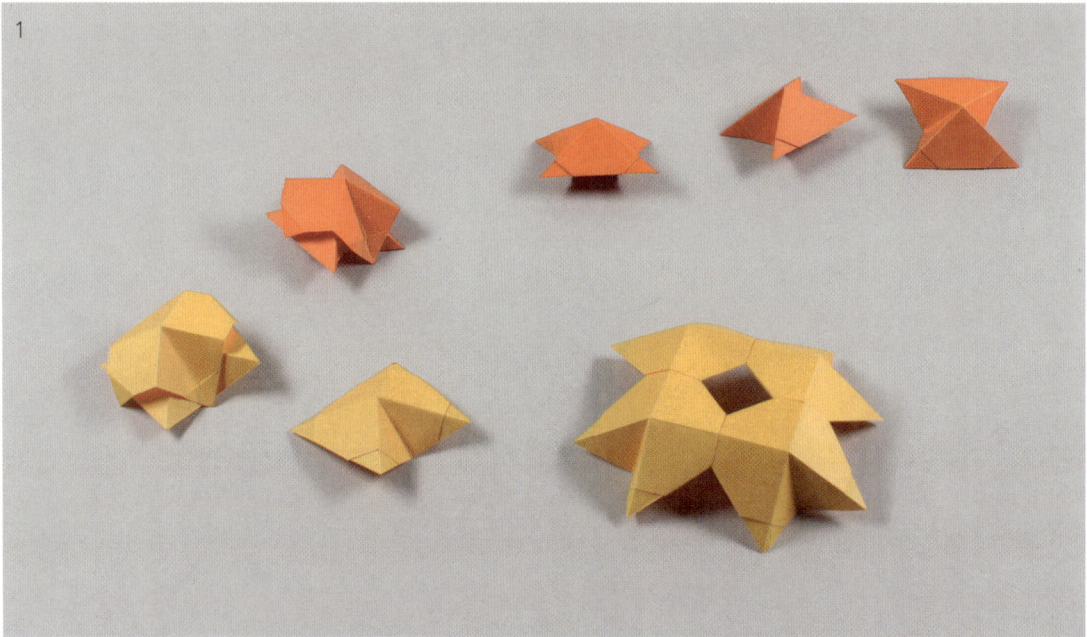

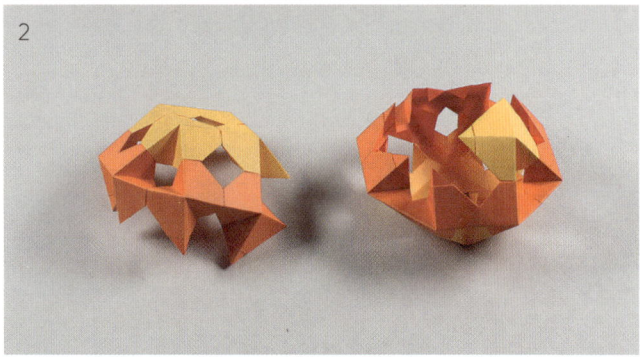

Twisted Rhombic Cubo-octahedron Cap

From Base to Module

You will need a total of 21 squares consisting of 8 × horse and 8 × rider in yellow, as well as 8 × horse and 8 × rider in orange—both bases of the same size. Fold mountain and valley folds according to the illustrations on the right. Please note: the yellow and orange bases have different mountain and valley folds. Join horse and rider to form modules in one color (fig. 1, *left and above*).

From Module to Finished Model

Join the first four yellow modules to form a quadrinomial ring joint (fig. 1, *right*) and repeat this step. Then join the orange modules—added to yellow modules—to form an orange band. Fig. 2 shows the progress from two sides. Finally, add a second yellow quadrinomial ring to the orange modules; this completes the model (fig. 3).

The resulting model looks like two diametrically twisted, joined caps or bottoms of the rhombic cubo-octahedron (model A03; p. 42). Fig. 4, *right*, shows the finished model next to the "rhombic cubo-octahedron cap" already introduced (see p. 128).

Difficulty: Easy

Simple cutting, simple construction; time required: ±1 hour

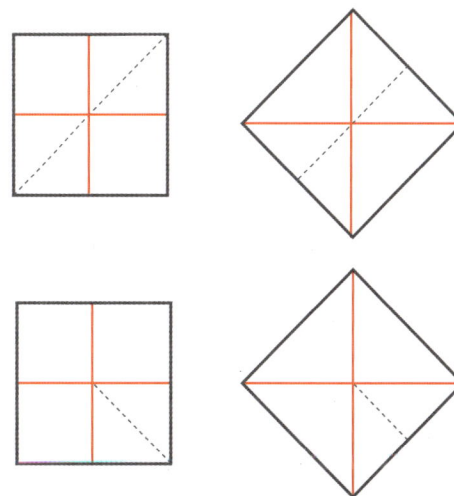

8 x   8 x   8 x   8 x

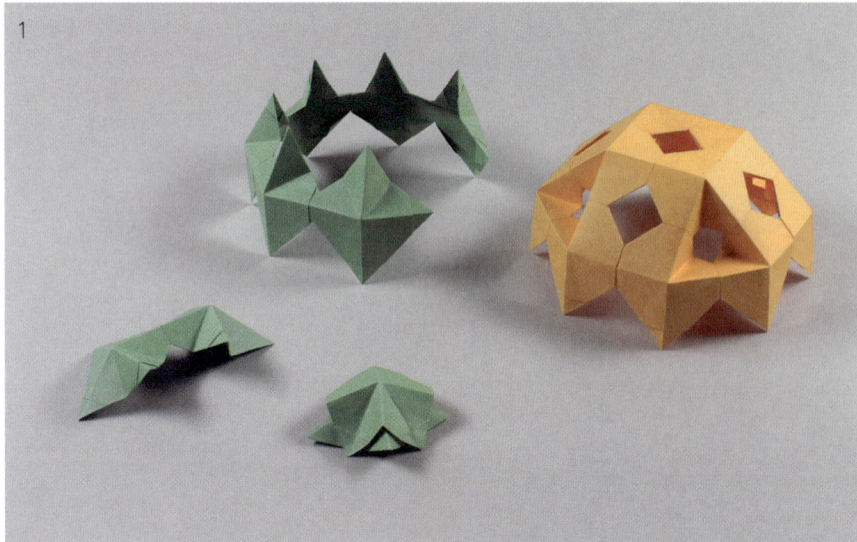

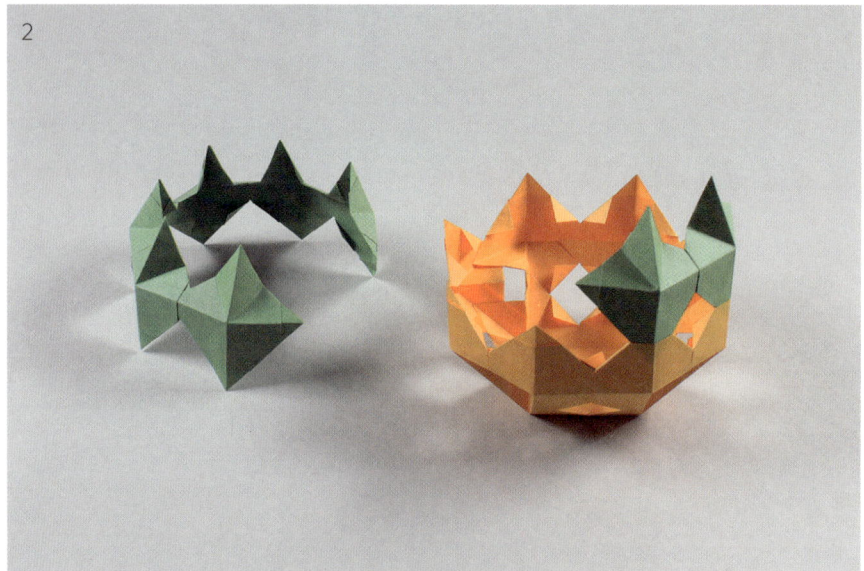

Rhombic Cubo-octahedron, Torso (3/4/3**)

First Steps

You will need a total of 48 squares, consisting of 12 × horse and 12 × rider in yellow, as well as 12 × horse and 12 × rider in green—both bases of the same size. Fold mountain and valley folds according to the illustrations on the right, and join horse and rider to form modules. Then join the yellow modules to form trinomial ring joints. Four trinomial rings form one-half of the polyhedron (fig. 1).

Putting the Model Together

Add the green modules—joining them up with yellow modules—to form a green band (figs. 2 and 3). The resulting form is similar to that of the rhombic cubo-octahedron (model A03; see p. 42). The green tips imply an eightfold center. In this torso, however, the model is not completed.

Difficulty: Easy

Simple cutting, simple construction; time required: ±75 minutes

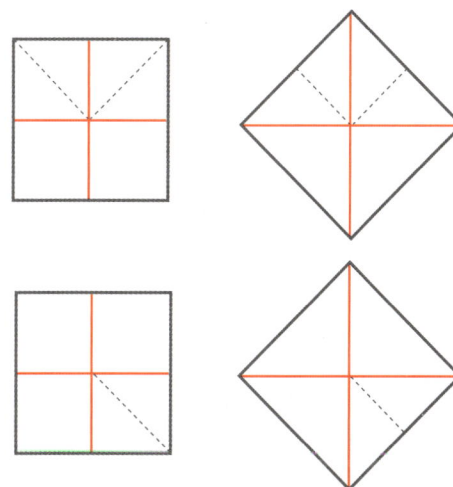

12 x   12 x   12 x   12 x

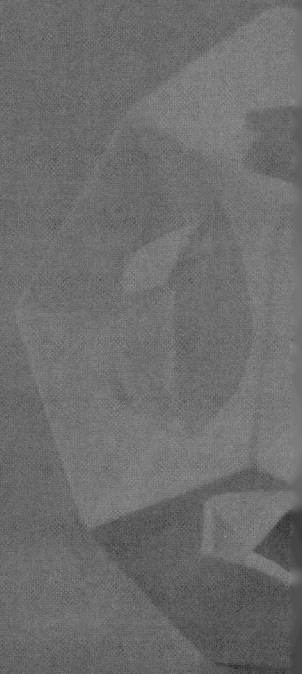

K-models
Extended Dodecahedra

At the heart of the models introduced in this chapter is the dodecahedron (model B04). Individual parts were taken from the original model and put into a different place, or extended with new parts. The colors chosen for the different models clearly illustrate the changes.

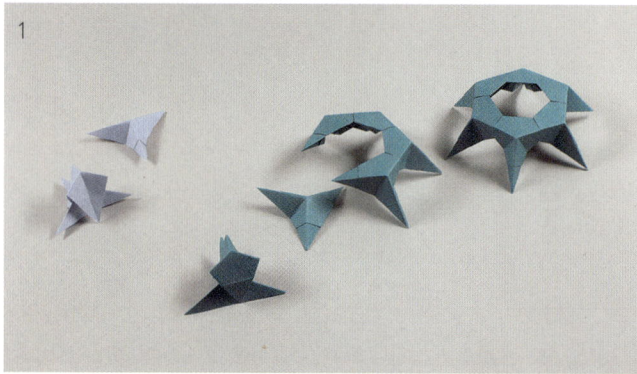

Extended Dodecahedron
Tetradecahedron (5/6)

From Base to Module

You will need 24 light-blue triangles (12 × horse and 12 × rider) and 24 turquoise triangles (12 × horse and 12 × rider)—in the same size and with the same folds. Fold mountain and valley folds according to the illustrations on the right. Then join horse and rider to form modules (fig. 1, *left*).

From Modules to Finished Model

Join threefold turquoise models to form two sixfold ring joints (fig. 1, *right*). Then add six trinomial light-blue modules at the tips of the first turquoise ring joint and complete with a further six light-blue modules (fig. 2, *left*). Finally, add the second turquoise ring joint (fig. 2, *right*; fig. 3). This completes the model.

Difficulty: Easy

Simple cutting, simple construction; time required: ±1 hour

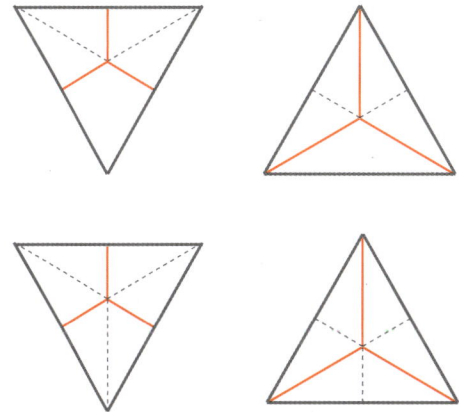

12 x   12 x   12 x   12 x

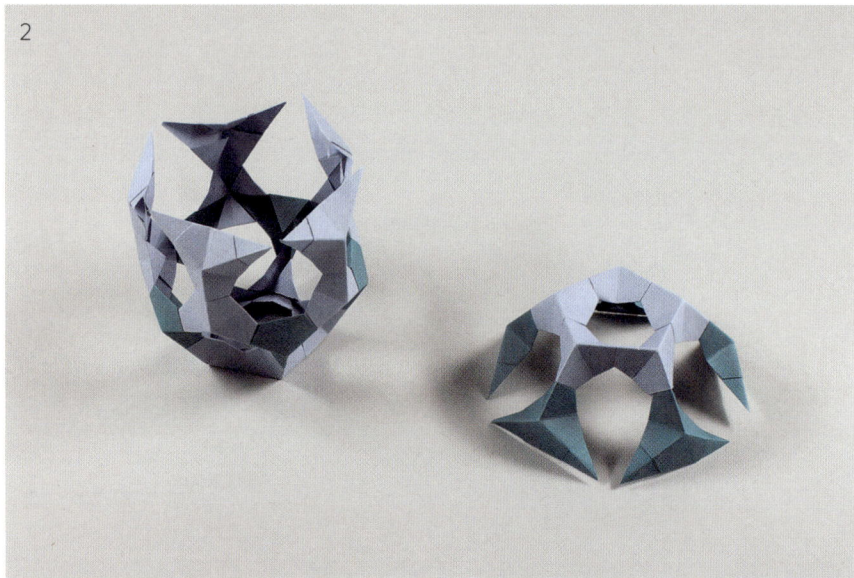

Extended Dodecahedron

Heptakaidecahedron (5/6')

From Bases to Modules

You will need a total of 40 light-blue triangles (20 × horse and 20 × rider) and 20 turquoise triangles (10 × horse and 10 × rider)—each of the same size. Fold mountain and valley folds according to the illustrations on the right. Then join horse and rider to form modules in one color.

From Modules to Model

Join the light-blue modules to form pentamerous ring joints (fig. 1, *right*) and add five turquoise modules (this constitutes one-half of the model; figs. 2 and 3). Repeat this step. Finally, add two light-blue modules each as a link between the two halves of the model. Figs. 4 and 5 show the finished model.

Difficulty: Medium

Cutting: first challenges; simple construction; time required: ±75 minutes

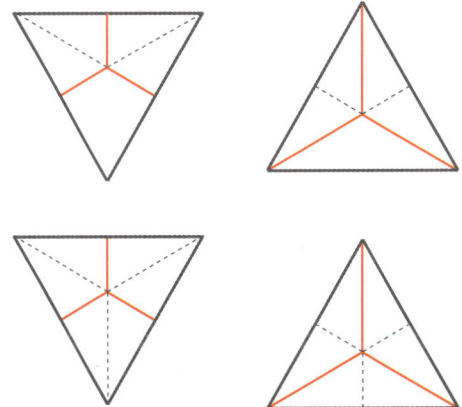

20 x   20 x   10 x   10 x

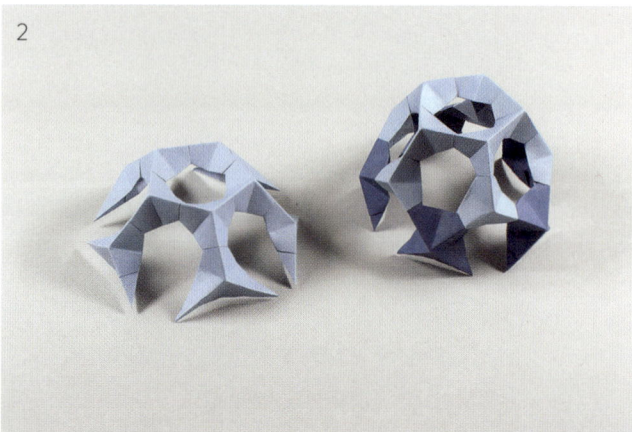

Extended Dodecahedron
Heptakaidecahedron (5/5*/4*)

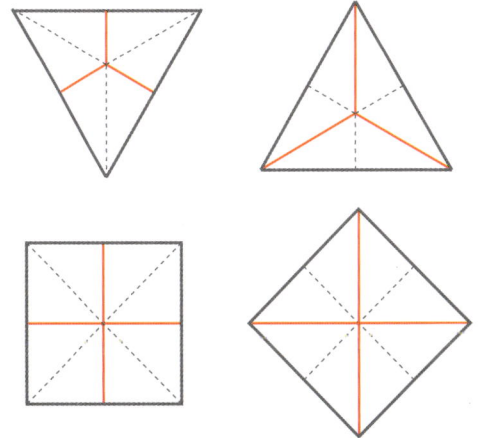

From Bases to Modules

You will need a total of 40 light-blue triangles (20 × horse and 20 × rider) and 10 dark-blue squares (5 × horse and 5 × rider)—each of the same size. Fold mountain and valley folds according to the illustrations on the right. Then join horse and rider to form modules.

From Modules to Model

Add five light-blue modules to form a first pentamerous ring joint (fig. 1, *right*; as in dodecahedron, model B04; see p. 46, fig. 1, *left*). Add another five light-blue modules and repeat this step (fig. 2, *left*). Then add dark-blue quadrinomial modules as a link between the two halves of the model (fig. 2, *right*). Finally, add the other pentamerous ring joint and complete the model (fig. 3, *right*). The finished model is very similar to the dodecahedron (see p. 46, fig. 3, *left*).

Difficulty: Medium

Cutting: first challenges; simple construction; time required: ±75 minutes

20 x   20 x   5 x   5 x

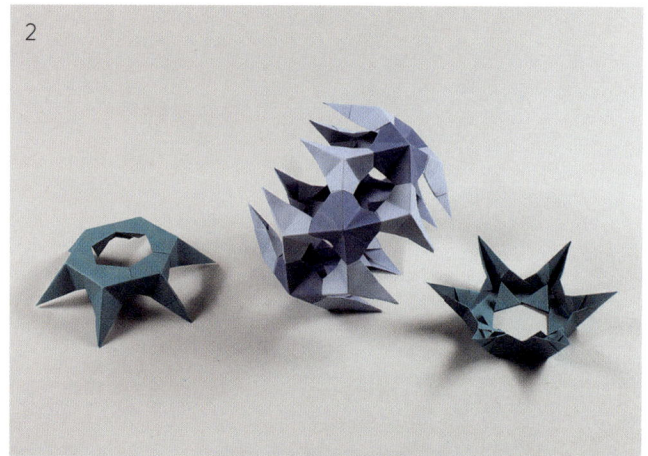

Extended Dodecahedron

Icosahedron (4*/5*/6)

From Base to Module

You will need 24 light-blue triangles (12 × horse and 12 × rider), 24 turquoise triangles (12 × horse and 12 × rider), and 12 dark-blue squares (6 × horse and 6 × rider)—each of the same size. Fold mountain and valley folds according to the illustrations on the right. Then join horse and rider to form modules (fig. 1, *left*).

From Module to Model

Join the light and dark blue modules to form a sixfold band (fig. 1, *right*) and join the turquoise models to form two sixfold rings (fig. 2). Finally, add the turquoise rings one after the other to the blue band (fig. 3). Figs. 4 to 6 show the finished model from different points of view.

Difficulty: Easy

Simple cutting, simple construction; time required: ±75 minutes

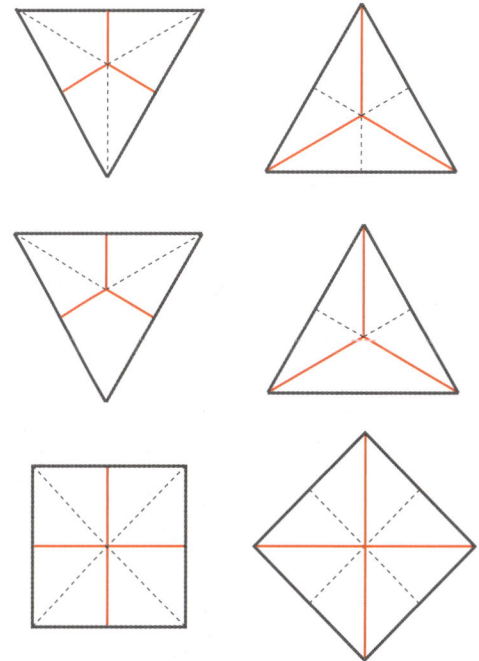

12 x	12 x	12 x	12 x	6 x	6 x

L-models
Extensions of Truncated Tetrahedra, Octahedra, and Icosahedra

We can meaningfully change the three forms of truncated tetrahedron (model B05; see p. 48), truncated octahedron (model B06; see p. 50), and truncated icosahedron (model B07; see p. 52) in the same way by adding link modules. This results in the respective twin form. Alternatively, you can join several truncated tetrahedra to form a chain.

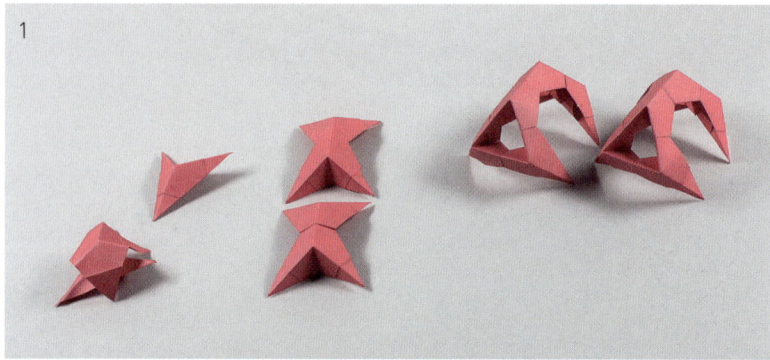

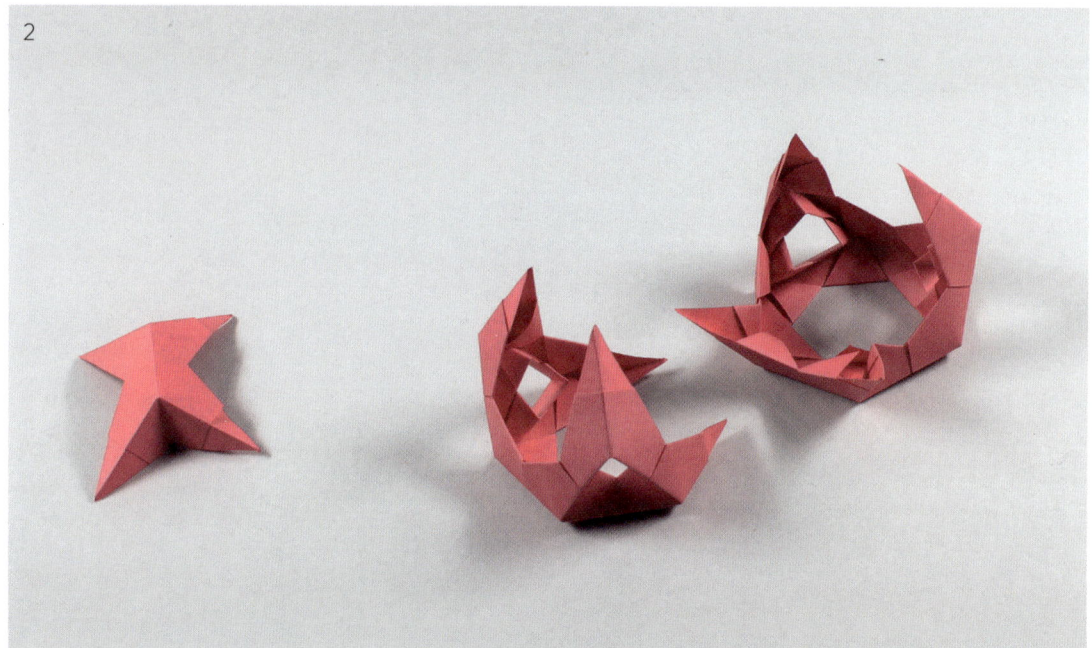

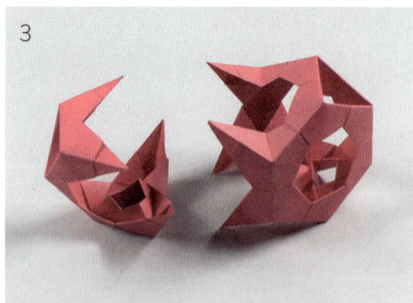

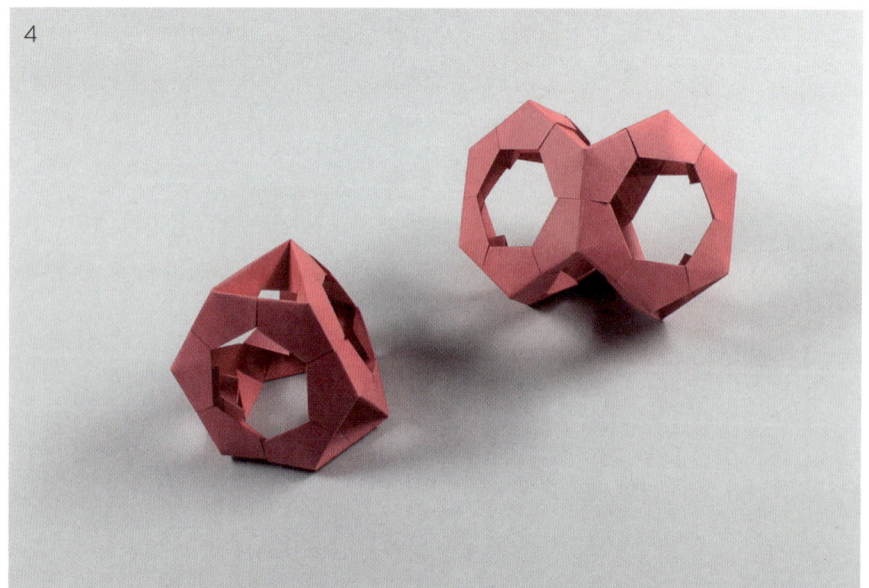

Twin Truncated Tetrahedron
Decahedron (3/6/6')

First Steps

You will need a total of 32 pink triangles of the same size, consisting of 16 × horse and 16 × rider. Fold mountain and valley folds according to the illustrations on the right. Then join horse and rider to form modules (fig. 1, *left*) and join three of these each (fig. 1, *center*). Join the remaining modules to form trinomial ring joints (fig. 1, *right*).

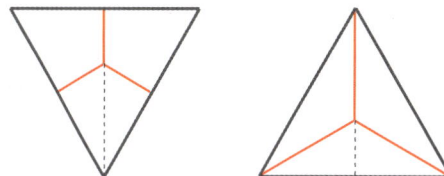

Constructing the Model

Join two trinomial ring joints each (figs. 1 and 2). Add the double modules (fig. 3, *right*) and complete the two model halves from fig. 3 to form the finished model (fig. 4, *right*). Fig. 4 shows the truncated tetrahedron (model B05; see p. 48) on the left.

Difficulty: Easy

Simple cutting, simple construction; time required: ±75 minutes

16 x   16 x

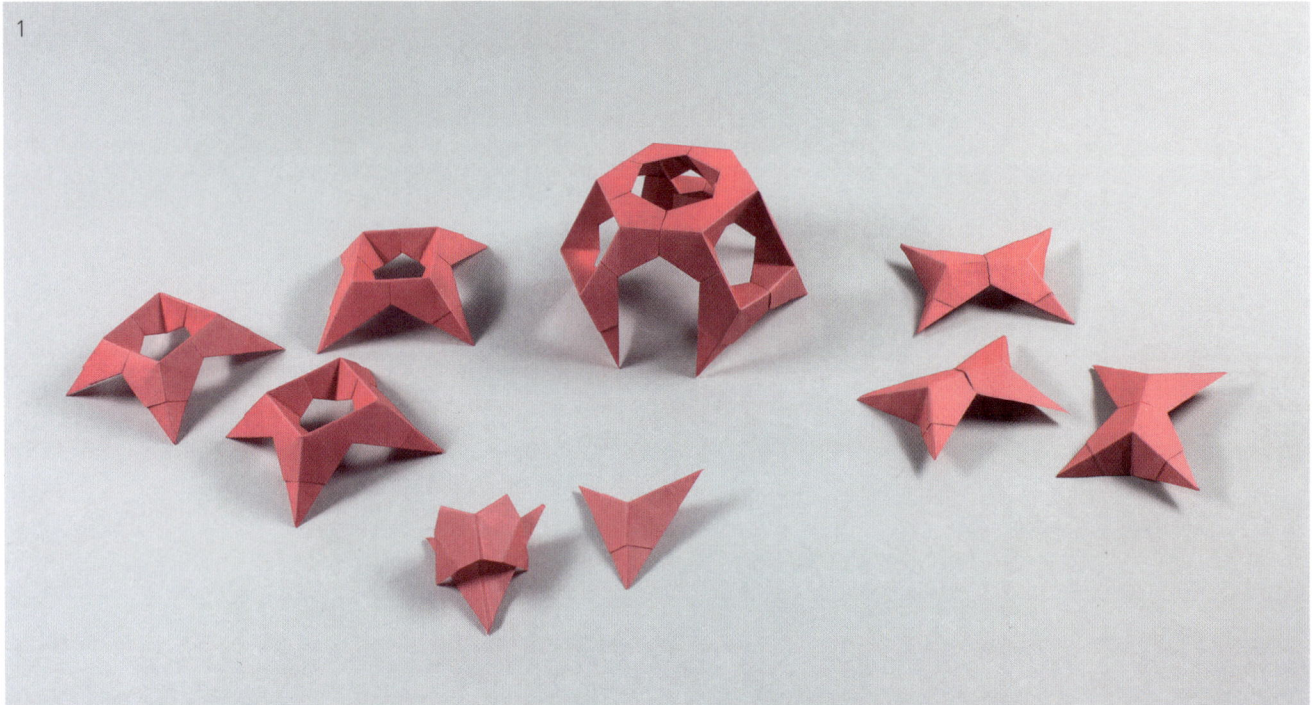

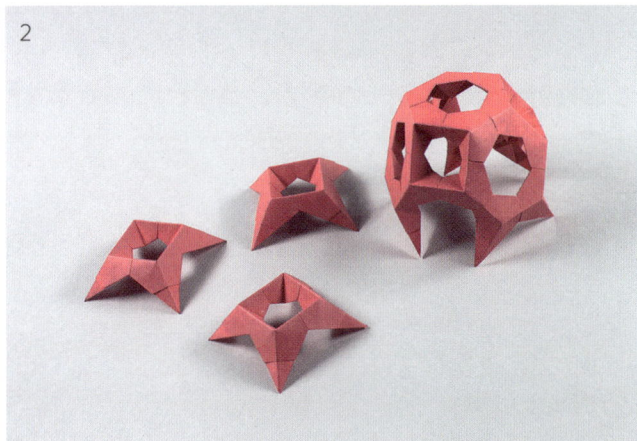

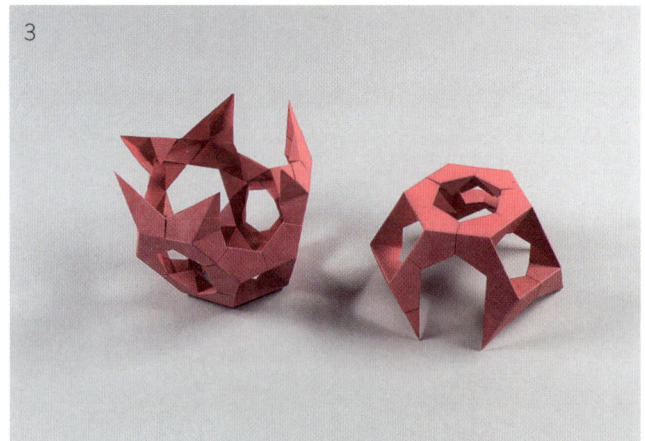

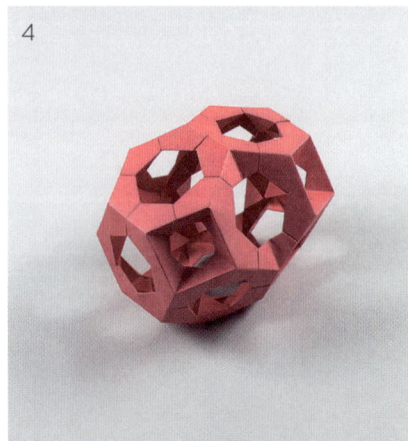

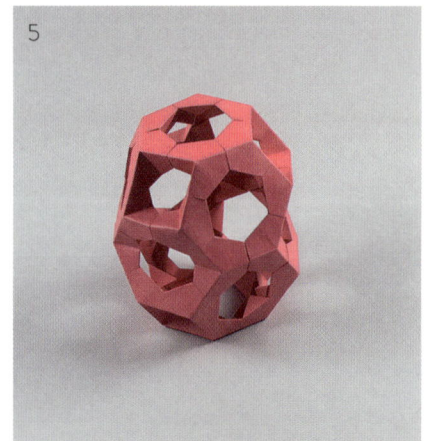

Twin Truncated Octahedron
Pentakaidecahedron (4/6/6')

First Steps

You will need a total of 60 pink triangles of the same size, consisting of 30 × horse and 30 × rider. Fold mountain and valley folds according to the illustrations on the right. Then join horse and rider to form modules in one color. Create six quadrinomial pink ring joints and put these together (fig. 1, *left*), then form two double modules (fig. 1, *right*).

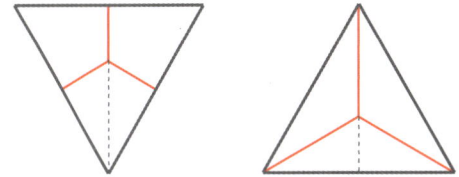

Constructing the Model

Join three quadrinomial ring joints each (fig. 2, *center*). This creates two halves of the model (fig. 3). Join these using the double modules from fig. 2 (fig. 4). The finished model is a twin model of the truncated octahedron (model B06; see p. 50). Figs. 4 and 5 show the model from different perspectives.

Difficulty: Easy

Simple cutting, simple construction; time required: ±1½ hours

30 x   30 x

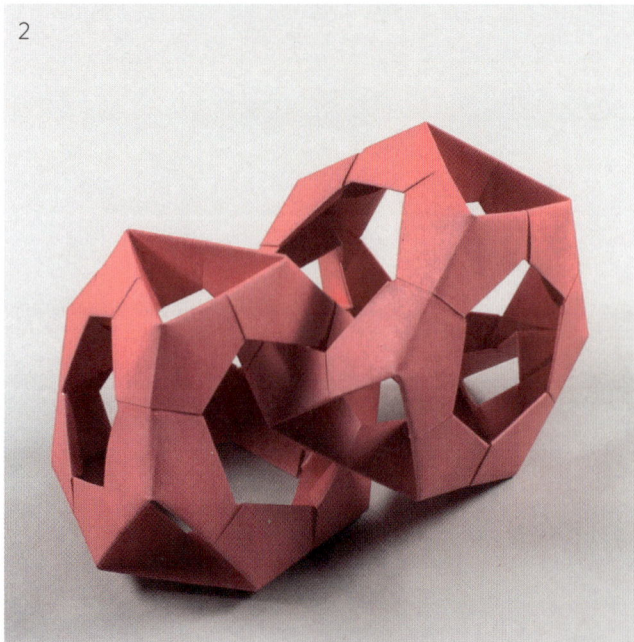

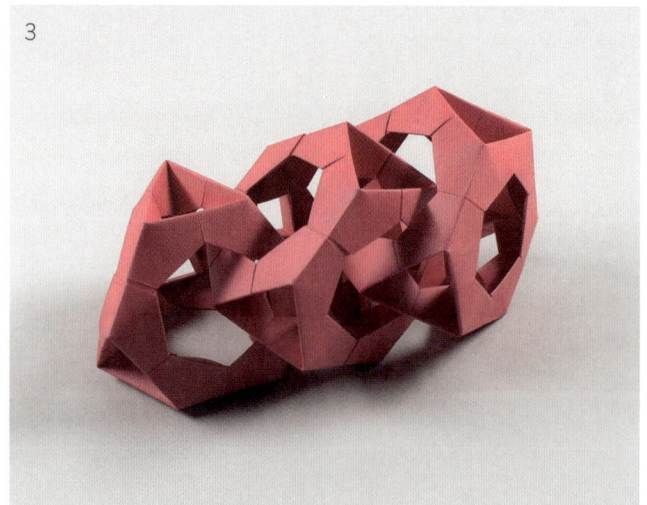

Penetrations: Truncated Tetrahedron
Octahedron

First Steps

You will need a total of 72 pink triangles of the same size, consisting of 36 × horse and 36 × rider (12 × horse and 12 × rider each per "chain member"). Fold mountain and valley folds according to the illustrations on the right. Then join horse and rider to form modules of one color and continue to join to form four trinomial ring joints (fig. 1). Then join the ring joints with each other. This model is identical in structure with the truncated tetrahedron (model B05; see p. 48).

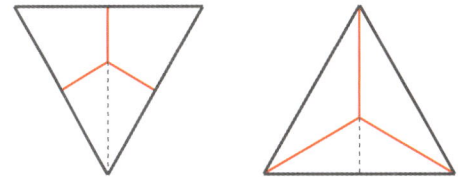

Completing the Model

Create modules and trinomial ring joints for a second model, as described. The link between the first two ring joints is achieved by a gap in the first model (fig. 1). The other two ring joints are added on the outside (fig. 2). This step can be repeated for any number of "chain members." Fig. 3 shows the finished model with three "chain members."

Difficulty: Easy

Simple cutting, simple construction; time required: ±1 hour

| 12 x | 12 x | 12 x | 12 x | 12 x | 12 x |

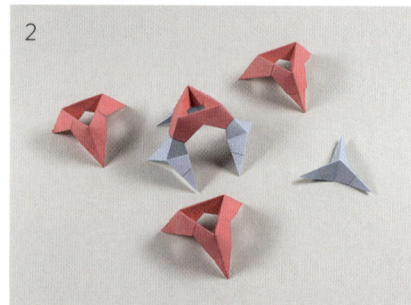

Extended Truncated Tetrahedron
Dodecahedron (3/6')

First Steps

You will need 24 pink triangles consisting of 12 × horse and 12 × rider, as well as 8 light-blue triangles consisting of 4 × horse and 4 × rider—all of the same size. Fold mountain and valley folds according to the illustrations on the right. Then join horse and rider into modules of one color and put them together to form four trinomial pink ring joints (fig. 1).

Completing the Model

Add individual light-blue modules to the first pink ring joint (fig. 2). Then add the next pink ring joint (fig. 3). Continue and complete the model in the same manner (fig. 4). The finished model (fig. 5) is an extended truncated tetrahedron (model B05; see p. 48).

Difficulty: Easy

Simple cutting, simple construction; time required: ±1 hour

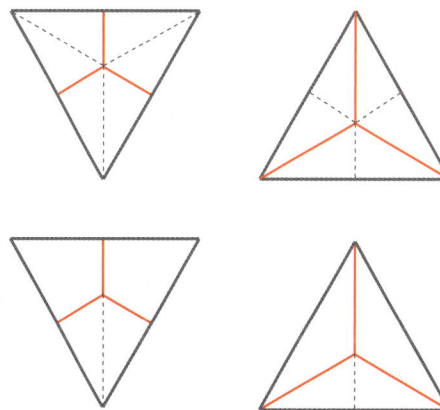

12 x   12 x   4 x   4 x

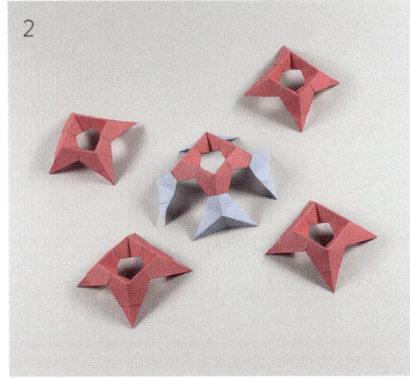

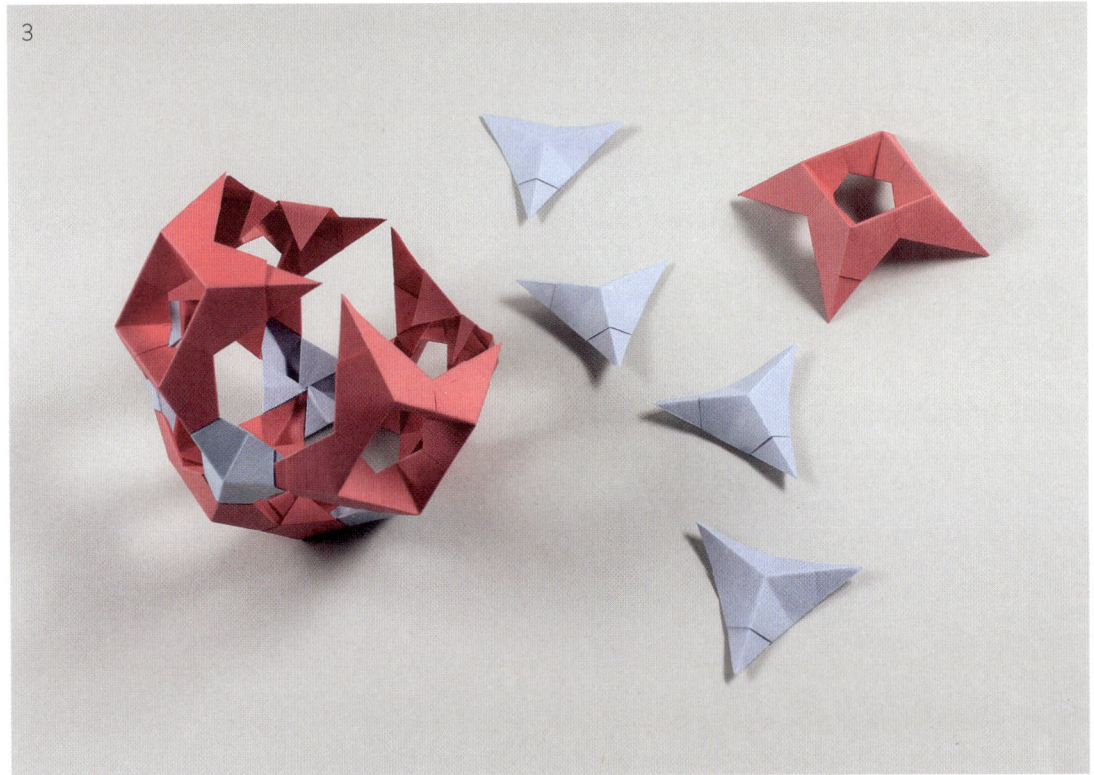

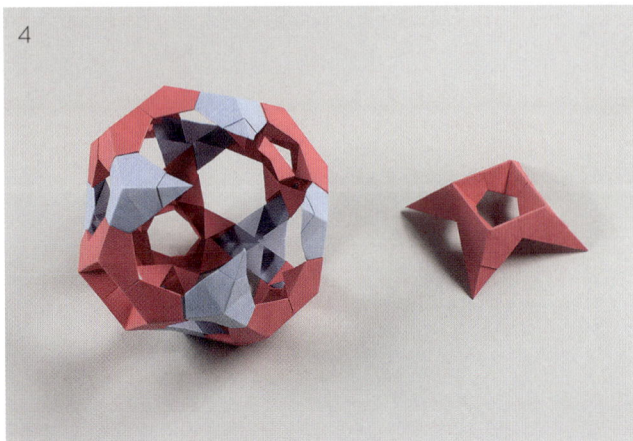

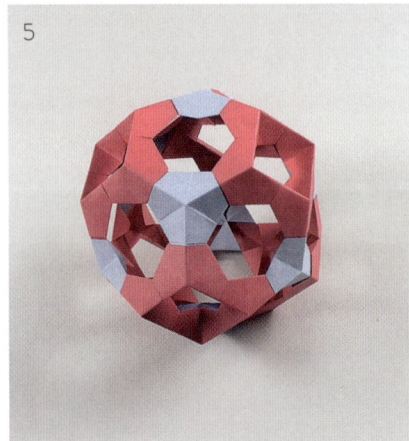

Extended Truncated Octahedron
Octakaidecahedron (4/6')

First Steps

You will need a total of 48 pink triangles consisting of 24 × horse and 24 × rider, as well as 16 light-blue triangles consisting of 8 × horse and 8 × rider—all of the same size. Fold mountain and valley folds according to the illustrations on the right. Then join horse and rider to form modules of one color and then join these to form six quadrinomial ring joints (fig. 1).

Further Construction

Add a light-blue module to the first pink ring joint (fig. 2). Then add the next pink ring joint (fig. 3). Continue and complete the model in the same manner (fig. 4). The completed model (fig. 5) results in an extended truncated octahedron (model B04; see p. 50).

Difficulty: Easy

Simple cutting, simple construction; time required: ±1½ hours

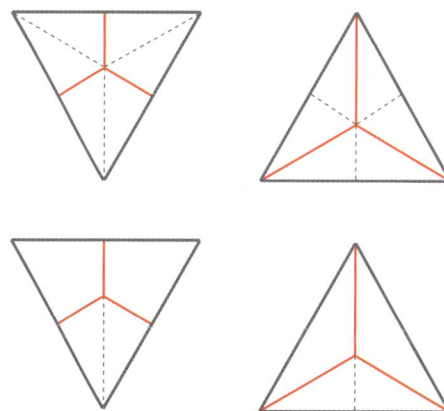

24 x   24 x   8 x   8 x

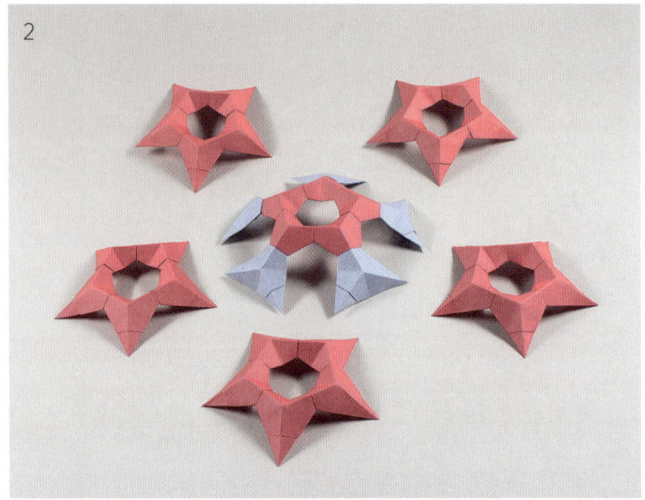

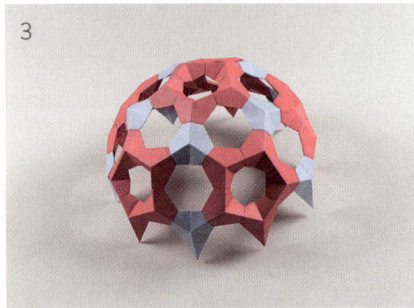

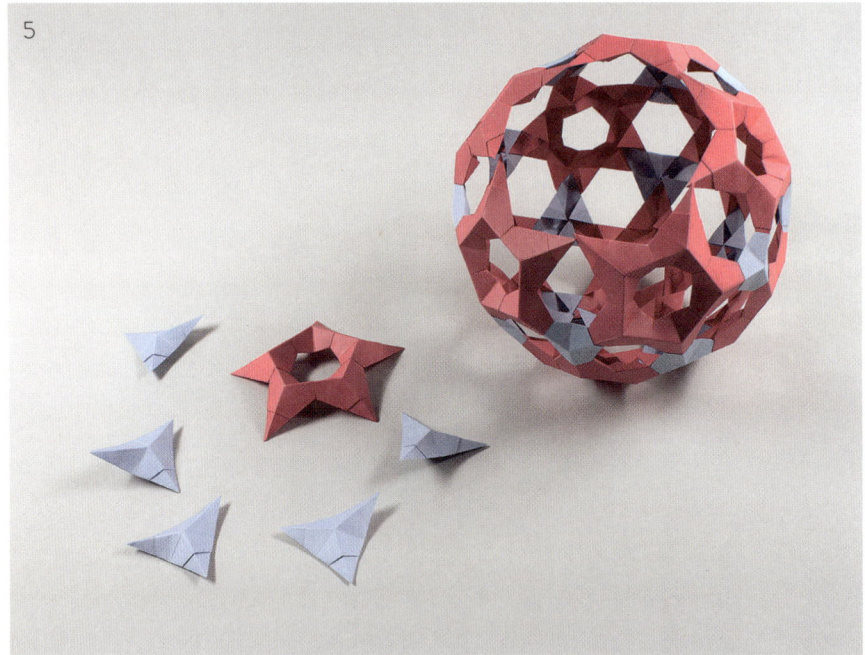

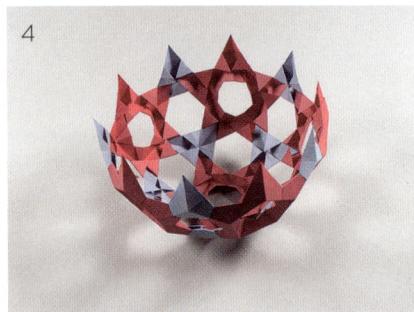

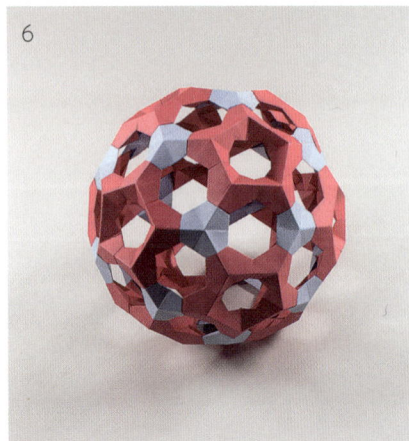

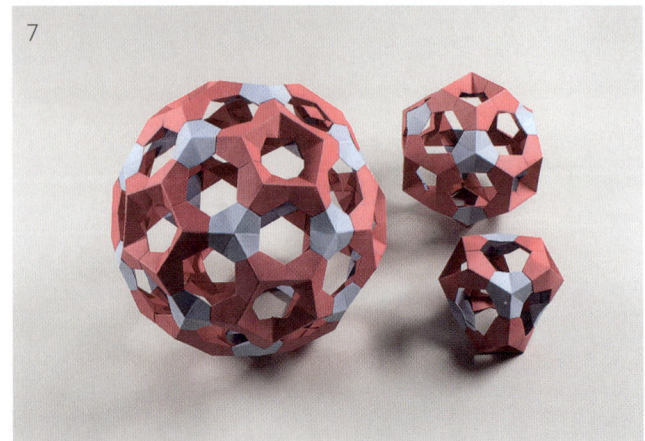

Extended Truncated Icosahedron
Tetracontadihedron (5/6*)

First Steps

You will need a total of 120 pink triangles consisting of 60 × horse and 60 × rider, as well as 40 light-blue triangles consisting of 20 × horse and 20 × rider—all of the same size. Fold mountain and valley folds according to the illustrations on the right. Then join horse and rider to form modules of the same size and put them together to form twelve pentamerous pink ring joints (fig. 1).

Further Construction

Add individual light-blue modules to the first pink ring joint (fig. 2). Then add the next five pink ring joints (fig. 3). Continue and complete the model in the same manner (figs. 4 and 5). The finished model (fig. 6) results in an extended truncated icosahedron (model B07; see p. 52). Fig. 7 also shows the two preceding models (see p. 152 and following).

Difficulty: Easy

Simple cutting, simple construction; time required: ±3½ hours

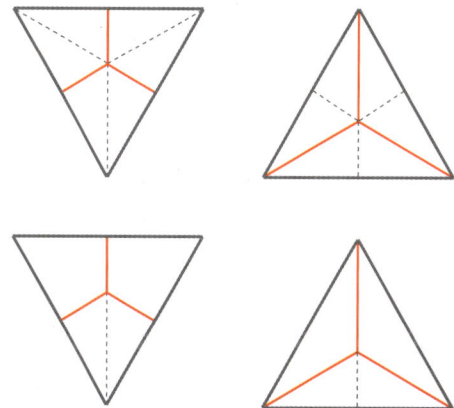

60 x   60 x   20 x   20 x

M-models
Extensions of Octahedron, Icosahedron, and Rhombic Dodecahedron

In this chapter, three forms serve as a point of departure for further variations. The squares of the octahedron (model A01; see p. 38) are partially substituted by pentagons and hexagons. Additionally, the pentagons of the icosahedron (model B08; see p. 54) are partially replaced with squares and hexagons, and the squares of the rhombic dodecahedron (model C09; see p. 58) with pentagons and hexagons.

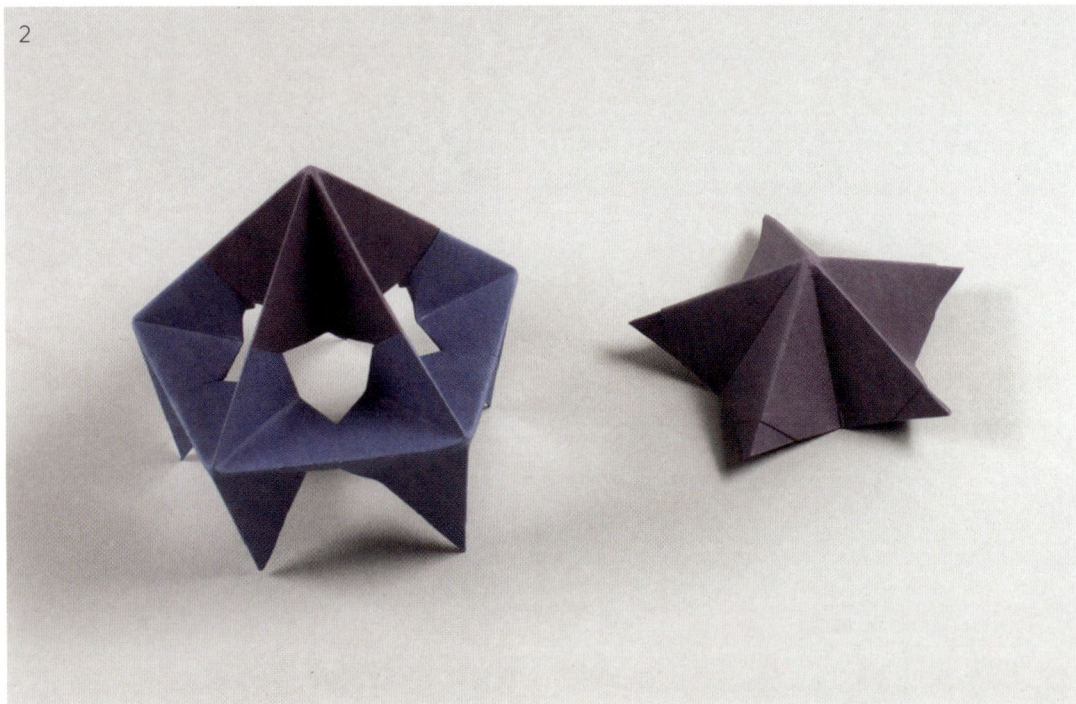

Modified Octahedron
Decahedron (3*)

From Bases to Modules

You will need a total of 10 dark-blue squares consisting of 5 × horse and 5 × rider, as well as 4 violet pentagons consisting of 2 × horse and 2 × rider—all of the same size and all with the same type of folds for each color. Fold mountain and valley folds according to the illustrations on the right. Then join horse and rider to form modules of the same color (fig. 1).

From Modules to Finished Model

Add only dark-blue modules to a first violet module (fig. 2). Join all five dark-blue tips with the second violet module (fig. 3, *right*). The finished pentamerous model resembles a quadrinomial octahedron (fig. 3, *left*; model A01; see p. 38).

Difficulty: Easy

Simple cutting, simple construction; time required: ±1 hour

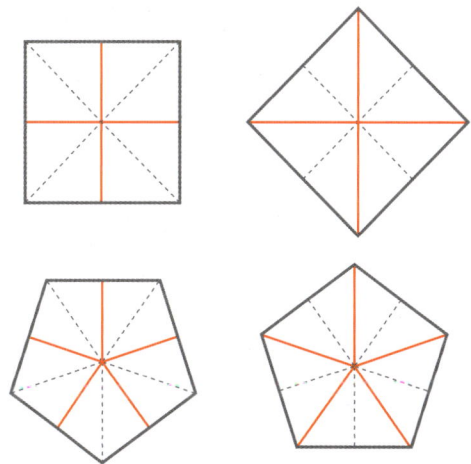

5 x   5 x   2 x   2 x

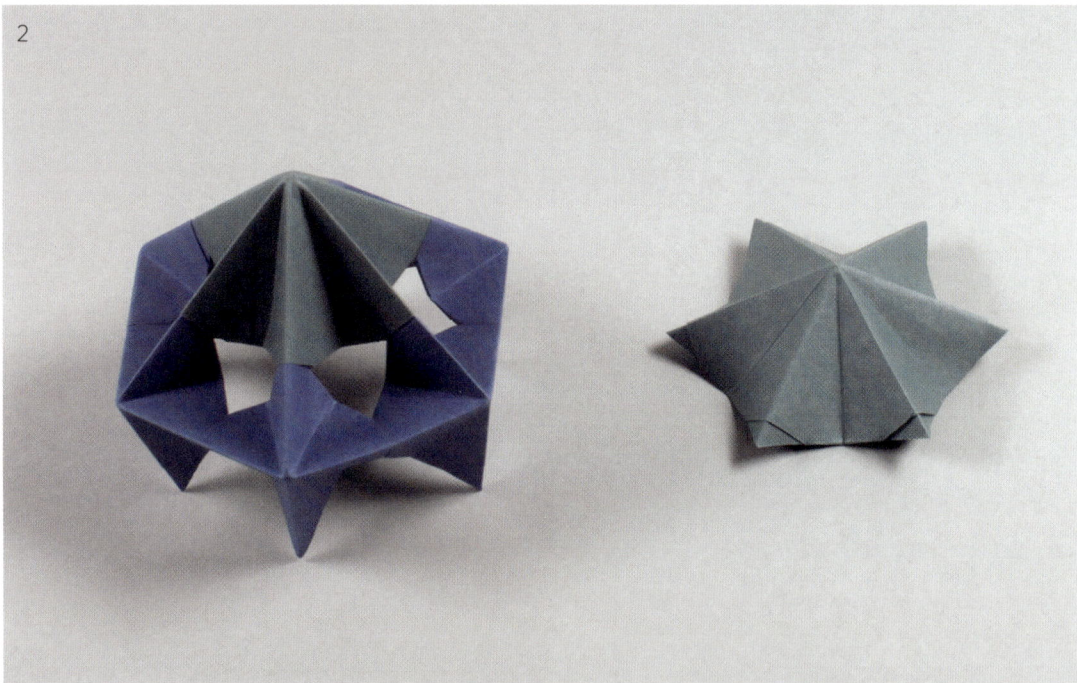

Modified Octahedron
Dodecahedron (3*)

From Bases to Modules

You will need 12 dark-blue squares consisting of 6 × horse and 6 × rider, as well as 4 moss-green hexagons consisting of 2 × horse and 2 × rider—all of the same size and the same folds for each color. Fold mountain and valley folds according to the illustrations on the right. Then join horse and rider to form modules in one color.

From Modules to Finished Model

Add only dark-blue modules to a first moss-green module (fig. 2). Join all six dark-blue tips with the second moss-green module (fig. 2, *right*). The finished model resembles a tetra-merous octahedron (fig. 3, *left*; model A01; see p. 38), as well as the preceding pentamerous model (fig. 3, *center*). Fig. 4 shows all three models from a different perspective.

Difficulty: Easy

Simple cutting, simple construction; time required: ±1 hour

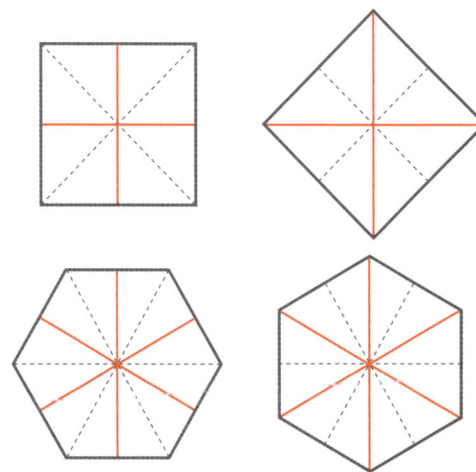

6 x   6 x   2 x   2 x

Modified Icosahedron
Hexakaidecahedron (3/3*)

From Bases to Modules

You will need 4 dark-blue squares consisting of 2 × horse and 2 × rider, as well as 16 violet pentagons consisting of 8 × horse and 8 × rider—all of the same size and with the same folds. Fold mountain and valley folds according to the illustrations on the right. Then join horse and rider to form modules (fig. 1).

From Modules to Finished Model

Join a dark-blue module with four violet modules. This results in the first half of the model (fig. 2, *left*). Build the second half in the same manner (fig. 3). Fig. 3 shows the relationship between this model and the icosahedron (model B08; see p. 54).

Difficulty: Medium

Cutting: first challenges; simple construction; time required: ±75 minutes

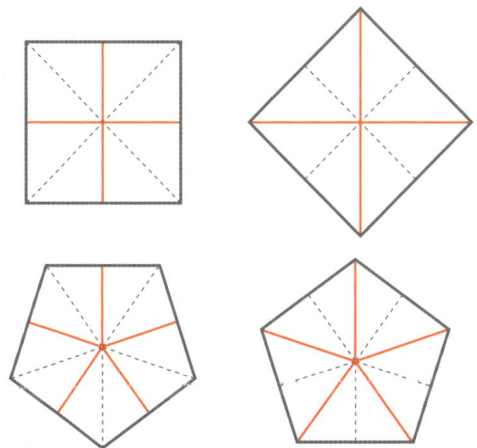

Modified Icosahedron (3/3*)

From Bases to Modules

You will need 24 violet pentagons consisting of 12 × horse and 12 × rider, as well as 4 moss-green hexagons consisting of 2 × horse and 2 × rider—in the same size, as well as with the same folds for each color. Fold mountain and valley folds according to the illustrations on the right. Then join horse and rider to form modules in one color (fig. 1).

From Modules to Finished Model

Join the violet modules to form a twelve-part band (figs. 2 and 3) and join the six ends at each side with the six ends of the moss-green sixfold module. Figs. 4 to 6 show the modified icosahedron (hexakaidecahedron, model M03; see p. 164), the icosahedron (model B08; see p. 54), and the modified rhombic dodecahedron (octakaidecahedron, model M05; see p. 168).

Difficulty: Medium

Cutting: first challenges; simple construction; time required: ±75 minutes

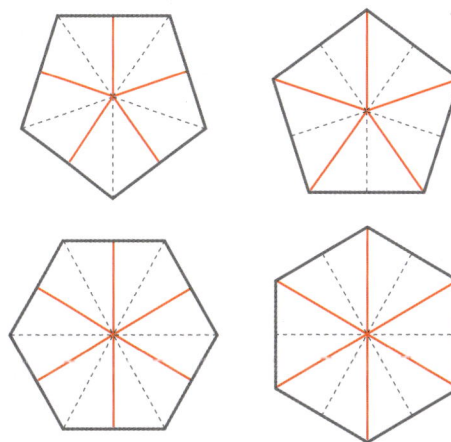

12 x   12 x   2 x   2 x

Modified Rhombic Dodecahedron
Octakaidecahedron (4*/4*)

From Bases to Modules

You will need 20 light-blue triangles (10 × horse and 10 × rider) and 10 dark-blue squares (5 × horse and 5 × rider)—in the same size and with the same folds. Fold mountain and valley folds according to the illustrations on the right. Then join horse and rider to form modules in one color (fig. 1, *below*).

From Modules to Finished Model

Join light-blue triangular and dark-blue square modules to form a pentamerous band (fig. 1, *above right*). Then add one violet module to one side and a second to the other (figs. 2 and 3). Figs. 4 and 5 show a model from different perspectives. The relationship between this model with the rhombic dodecahedron (model C09; see p. 58) can be clearly seen in fig. 1 (*above left*).

Difficulty: Medium

Cutting: first challenges; simple construction; time required: ±75 minutes

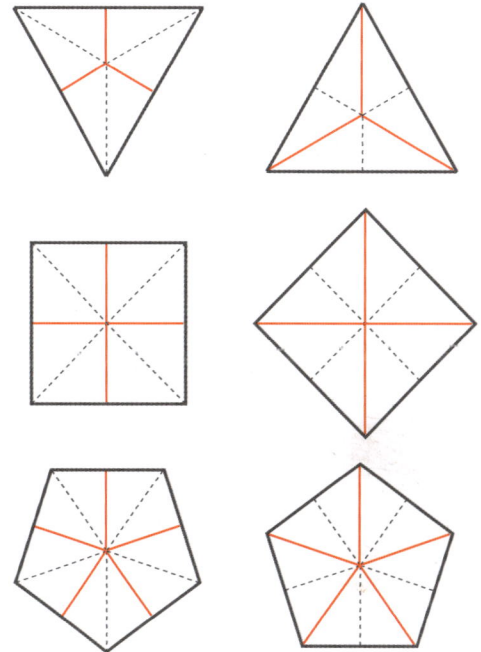

| 10 x | 10 x | 5 x | 5 x | 2 x | 2 x |

Modified Rhombic Dodecahedron
Icosatetrahedron (4*/4*)

From Bases to Modules

You will need 24 light-blue triangles (12 × horse and 12 × rider), 12 dark-blue squares (6 × horse and 6 × rider), and 4 moss-green hexagons (2 × horse and 2 × rider)—all of the same size and with the same folds for each color. Fold mountain and valley folds according to the illustrations on the right. Then join horse and rider to form modules in one color.

From Modules to Finished Model

Join the light-blue ternary modules with dark-blue tetramerous modules to form a continuous band (fig. 1, *above right*). Then join a sixfold module at each of the six light-blue tips on each side (figs. 2 and 3). Figs. 4 and 5 show two different views of the finished model.

Difficulty: Medium

Cutting: first challenges; simple construction; time required: ±75 minutes

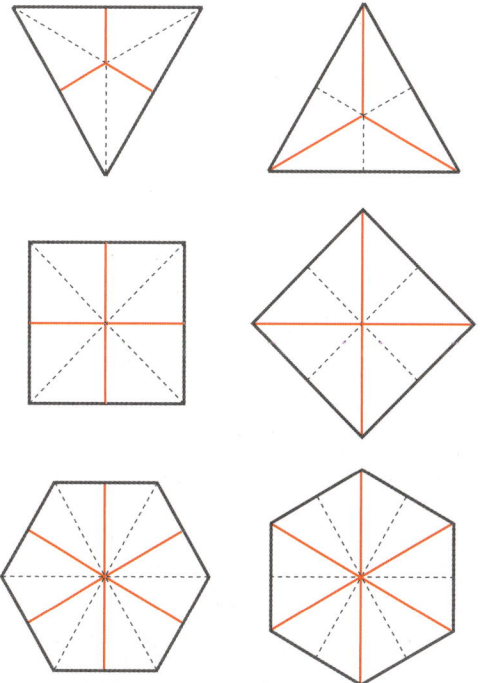

| 12 x | 12 x | 6 x | 6 x | 2 x | 2 x |

Appendix

Afterword

Why did I want to realize an entire sequence of geometrical models following a particular type of construction principle? In the beginning was the octahedron—built using cheap paper that was ill suited to the purpose. And in these first attempts, further forms developed. The pleasure in beautiful forms and their mysterious, lawful relationships among each other eventually created the occasion for daring to approach other forms as well. Initial technical challenges were overcome with good ideas. In the end, it was the striving for completion that drove me to realize all regular and semiregular polyhedra.

At the same time, opportunities arose to give little presentations, seminars, and workshops, where it became clear that folded polyhedra are a rewarding subject: the folding is very easy and can be done with nearly no previous experience, and it is a repetitive task that allows the mind to roam freely. With the finished spatial model, you have an aesthetic "added value" that is more than the sum of the two-dimensional individual parts. Furthermore, you can even continue working on it on the train journey home and experiment with it on your own. Then my first articles on folded polyhedra as a confluence of geometry, craft, and art were published. There was even the Phänomena prize, funded by Friedhelm Kürpig, which was won in 2007 in

the context of a seminar of the Deutsche Gesellschaft für Geometrie und Grafik (DGfGG) in Kornelimünster, near Aachen. In the context of the last models, a range of difficulties appeared that practically required my utmost energies and an extreme time commitment. Finally, a happy coincidence and suitable circumstances enabled me to write a book about it all. I was surprised and delighted to find out that others also tried to build the sequence of forms treated in this book, and have reached very useful results too. Especially in the "gyroscope" models of Rona Gurkewitz and Bennett Arnstein (see bibliography, p. 178) some of these forms were realized. In their publications you will find the octahedron (model A01), dodecahedron (model B04), truncated tetrahedron (model B05), truncated octahedron (model B06), truncated icosahedron (model B07), rhombic dodecahedron (model C09), rhombic tricontahedron (model C10), triakis icosahedron (model C11), deltoidal icosatetrahedron (model D12), deltoidal hexecontahedron (model D13), pentagonal icosatetrahedron (model D14), triakis octahedron (model H26), pyramidal octahedron 4/6/8 (model H29), pentakis dodecahedron 3/10 (model H30), and pentakis dodecahedron 4/6/10 (model H31), as well as some of the extensions presented in chapter J.

In contrast to Gurkewitz/Arnstein, I have chosen a different, more intuitive approach in *Folding Polyhedra*. Additionally, in comparison with the method of realization of the models shown by Gurkewitz/Arnstein, I was able to make some small improvements. For example, some of the valley folds used by them can be left out (e.g., in models B05, B06, and B07). In this sense I am looking forward to seeing whether even with regard to my own discoveries, further improvements and extensions can be found in the future. Suggestions are very welcome and should be sent to the publisher's address. I cannot, however, guarantee an answer in each and every case.

In chapters J–M, I have presented some less regular, extended forms that still—at least at a second glance—have a clear relationship with the polyhedra forms. These forms are mostly very easy to build and form a relaxing counterpart to the more difficult and very difficult polyhedra forms of chapters E–H. The idea behind these "extensions" follows a new approach: here I have experimented with combining the same basic forms in new ways. Many more forms than I could include in this book were thus created: chains, cylindrical forms, prismatic forms, torsos, and entirely free forms. Under favorable circumstances these could potentially form the basis for a second volume. The diversity of forms realized in *Folding Polyhedra* is thus not at all exhausted yet.

To combine craft, art, and science as explicitly mentioned in the subtitle accommodates the wish that many people entertain: to work across the boundaries of disciplines, both practically and theoretically. For the practical realization of my manuscript, this sometimes created a complex challenge that asked for a mediatory approach on the levels of language, illustrations, and photography that deviates considerably from usual origami instructions.

The final result was possible only through the excellent collaboration among author, layout, publishing, and copy editors. We hope that we have successfully communicated a complex subject in such a way that its practical realization becomes easy to understand, and that during the actual construction the main focus can be on the experience of one's own abilities. May these grow with the tasks. We wish you success with everything you attempt! Because of the complexity of the subject matter, we have been especially careful to avoid any mistakes in the instructions and numbers. However, no liability can be accepted.

Glossary

Platonic solids: Synonym for regular polyhedra. The five regularly shaped three-dimensional forms designated after Plato (ca. 428–347 BCE) are tetrahedron (see folded polyhedra model E17 on p. 78), cube/hexahedron (see folded polyhedra model E16 on p. 76), octahedron (see folded polyhedra model A01 on p. 38), dodecahedron (see folded polyhedra model B04 on p. 38), and icosahedron (see folded polyhedra model B08 on p. 54). These were mentioned for the first time in Plato's writings (in *Timaeus* and *Phaedo*). See figure on facing page, *center*.

Archimedean solids: Thirteen semiregular polyhedra (plus two chiral forms) designated after the Greek mathematician Archimedes (ca. 287–212 BCE). Each Archimedean solid can be traced back to one or two Platonic solids, which gives us the name of the Archimedean solid. Additionally, each Archimedean solid has a Catalan dual partner (see **Catalan solids**). After Pappus of Alexandria, it was Archimedes who first described this sequence of forms in its entirety. In the "Elements" of Euclid—a mathematical treatise—we find the earliest known presentation of the Archimedean solids. See figure on facing page, left half of illustration.

Catalan solids: Thirteen semiregular polyhedra (plus two chiral forms) designated after the Belgian mathematician Eugène Charles Catalan (1814–1894). Each Catalan solid can be traced back to one or two Platonic solids, which gives us the name of the Catalan solid. Each Catalan solid has its Archimedean dual partner (see **Archimedean solids**). Around 2,000 years after Archimedes, Catalan was the first to describe this sequence of polyhedra in its entirety. See facing page, right half of illustration.

polygon. (Greek: *poly*, many; and *gon*, edge); generally, a solid with many edges or vertices. Example: square.

polyhedron: (Greek: *poly*, many; and *hedron*, area, surface); generally, a solid with many surfaces. Example: cube.

regular: Even or uniform. A polygon is regular when all sides have the same length and all angles between the edges are the same. A polyhedron is regular when all edges are of the same length, all surfaces are uniform, and the angles between neighboring surfaces are the same. All Platonic solids are regular polyhedra.

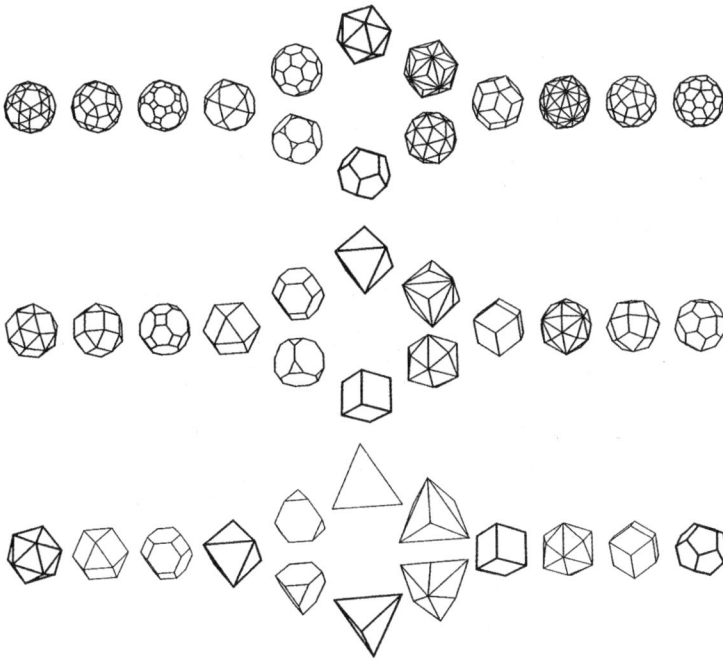

All polyhedra forms at a glance:
In the center: Platonic solids, *left half*: Archimedean solids, *right half*: Catalan solid (illustration by Helmut Emde, adjusted; see bibliography, p. 178)

semiregular: Only partially even or uniform. A polyhedron is semiregular when it consists of the same surfaces, but the angles at the vertices are of different sizes (Catalan solids). Alternatively it can consist of different types of surfaces, but in this case all edges have to have the same length (Archimedean solids).

chiral: Chiral three-dimensional forms can be constructed in two versions, each a mirror image of the other. They are like the right hand in relation to the left. Example: cubus simus (model F21).

dual partner: This can be described as an "opposite twin." Example of the twin characteristic: cube and octahedron have twelve edges each. Example of the opposite characteristic: the cube has six surfaces and eight vertices, while the octahedron conversely has eight surfaces and six vertices.

flat pattern: Spreading of the individual surfaces of the three-dimensional form next to one another in a flat surface. Example: Dürer's dodecahedron (see p. 12, fig. right). All twelve pentagons lie next to each other and are connected only through a single neighboring surface.

mesh gap: This designates the gap in the flat pattern between two edges of a polygon that in the three-dimensional model would form an edge of the polyhedron (see p. 26).

pyramidal polyhedron: Example: triakis octahedron, or pyramidal cube (model H26). Starting from a cube, the square faces are pulled out from the center of the square (like a tent). This results in a flat, four-faced pyramid on the face of the square.

truncated polyhedron: Example: truncated cube (model F18). If you cut back the edges of a cube by the same amount, this results in a three-sided cut surface, and the original surfaces of the cube are turned into octagonal "trunk" surfaces.

topology: (Greek: *tópos*, shape, form, "Gestalt"). The science of the shape of forms

Bibliography and List of Illustrations

BIBLIOGRAPHY

Adam, Paul, and Arnold Wyss. *Platonische und Archimedische Körper, ihre Sternformen und polaren Gebilde.* 2nd ed. Stuttgart: Verlag Freies Geistesleben, 1994.

Beutelspacher, Albrecht, and Laila Samuel, eds. *Ecken und Kanten—Europaweit: Drei Länder—Vier Künstler—150 Geometrische Meisterwerke.* Exhibition catalog Mathematicum e.V. Gießen, includes works by Friedhelm Kürpig, Ulrich Mikloweit, Rinus Roelofs, and Ueli Wittorf, April 12–May 11, 2014.

Bindel, Ernst. *Johannes Kepler: Mathematiker der Weltgeheimnisse; Beiträge zu seinem Lebensbild.* Stuttgart: Verlag Freies Geistesleben, 1971.

Catalan, Eugène Charles. "Mémoire sur la théorie des polyèdres." *Journal de l'École Impériale Polytechnique* 24 (1865): 1–71.

Dürer, Albrecht. *Unterweisung der Messung mit dem Zirkel und Richtscheit.* Facsimile edition of the Nuremberg 1525 edition. Nördlingen, Germany: Verlag Dr. Alfons Uhl, 2000.

Emde, Helmut. "Zur Geometrie räumlicher Strukturen." In *Diatomeen I: Schalen in Natur und Technik.* Edited by Klaus Bach and Berthold Burkhardt, 222–43. IL 28. Stuttgart: Institut für leichte Flächentragwerke der Universität Stuttgart, 1984.

Euclid. *Die Elemente.* Edited and translated by Clemens Thaer. Darmstadt: Wissenschaftliche Buchgesellschaft, 1975. [An English translation of *The Elements* by Thomas L. Heath (1905) is still in print.]

Goethe, Johann Wolfgang von. *Werke in zwei Bänden.* Vol. 2. Munich and Vienna: Carl Hanser Verlag, 1981. [An English translation published by Delphi Classics of the Complete Works is currently not in print, but available free as a virtual resource.]

Gurkewitz, Rona, and Bennett Arnstein. *Multimodular Origami Polyhedra: Archimedeans, Buckyballs and Duality.* Mineola, NY: Dover, 2003.

Heinz, Alexander. "Falt-Polyeder: Eine west-östliche Verbindung." *Informationsblätter der Geometrie,* February 2012: 22–27.

Heinz, Alexander. "Kulturgeschichte und geometrische Aspekte zur Entwicklung des Raumbewusstseins." *Mensch und Architektur* 60 (November 2007): 58–63.

Heinz, Alexander. "Das Runde muss ins Eckige: Ballformen und ihre Grundlagen." *Informationsblätter der Geometrie*, January 2017.

Jamnitzer, Wenzel. "Perspectiva." Corporum Regularium, Nuremberg, 1568.

Kepler, Johannes. *Weltharmonik*. Translated and introduced by Max Kaspar. Munich and Leipzig: R. Oldenbourg, 1939. [A partial translation into English is available as *Harmonies of the World*, Amazon UK.]

Kraul, Walter. *Platonische Körper und ihre Verwandlungen*. Stuttgart: Verlag Freies Geistesleben, 2014.

Kürpig, Friedhelm, and Koos Vorhoeff. *Round about—über Ecken und Kanten*. Exhibition catalog Mathematicum Gießen, April 6–May 5, 2019, in context of the series Modern Mathematical Art.

Marshall, Dorothy N. "Carved stone balls." In *Proceedings of the Society of Antiquaries of Scotland* 108 (1976): 40–72.

Melchizedek, Drunvalo. *Die Blume des Lebens*. Vol. 1. Burgrain, Germany: Ueberreuter Buchproduktion, 2004. [See pp. 155–184].

Pacioli, Luca. *Divine proportione*. Illustrations by Leonardo da Vinci. Reprint of the original *Divina proportione*, with French translation. Paris: Librairie du Compagnonnage, 1988.

Pauling, Linus, and Roger Hayward. *The Architecture of Molecules*. San Francisco: Freeman, 1964.

Plato. *Euthyphro, Apology, Crito, Phaedo*. Greek with translation by Chris Emlyn-Jones and William Preddy. Loeb Classical Library 36. Cambridge, MA: Harvard University Press, 2017.

Plato. *Timaeus and Critias*. English translation by Andrew Gregory. New York: Oxford University Press, 2008.

Schläfli, Ludwig. *Theorie der vielfachen Kontinuität*. Edited by H. J. Graf for the Denkschriften-Kommission der Schweizer Naturforschenden Gesellschaft. Zurich, Switzerland: Zürcher & Furrer, 1901.

Schlegel, Victor. *Theorie der homogen zusammengesetzten Raumgebilde*. Nova Acta Leopoldina 44, no. 4. Leipzig: W. Engelmann, 1883.

Schuré, Edouard. *Die großen Eingeweihten: Geheimlehren der Religionen*. Cologne: Anaconda, 2006. [English translation: *The Great Initiates: A Study of the Secret History of Religions*. New York: Steinerbooks, 1992]

Teichmann, Frank. *Der Mensch und sein Tempel.* 4 vols.: *Chartres: Schule und Kathedrale*; *Megalithkultur in Irland, England und der Bretagne*; *Ägypten*; and *Griechenland*. Stuttgart: Verlag Urachhaus, 1999–2003.

Williams, Robert. *The Geometrical Foundation of Natural Structure: A Source Book of Design.* Mineola, NY: Dover, 1979.

Ziegler, Renatus. *Platonische Körper: Verwandtschaften, Metamorphosen, Umstülpungen.* Dornach, Switzerland: Verlag am Goetheanum, 2008.

ILLUSTRATIONS

page 11
top
© Graham Challifour
www.ancient-wisdom.com/scotlandballs.htm

top right
© Alexander Heinz, own drawing after illustrations in Melchizedek 2004, pp. 161–63

page 12
left
© Wikimedia Commons, Kleon3, CC-SA-4.0 International

center
From Luca Pacioli, *Divina proportione*, plates XXVII and XXVVIII (Milan, 1509).

right
From Albrecht Dürer, *Unterweisung der Messung mit dem Zirkel und Richtscheit*, Viert Büchlein, plate 33 (Nuremberg, 1525).

page 13
left
From: Kepler, Johannes, *Harmonices Mundi*. Linz 1619.

center
From Johannes Kepler, *Mysterium Cosmographicum*, tabula III: Orbium planetarum dimensiones, et distantias per quinque regularia corpora geometrica exhibens (Tübingen, Germany, 1596).

right
From Eugène Charles Catalan, "Mémoire sur la théorie des polyèdres," *Journal de l'École Impériale Polytechnique* 24 (1865), plate V.

page 14
left
© Friedhelm Kürpig

right
From Victor Schlegel, *Theorie der homogen zusammengesetzten Raumgebilde* (Leipzig: W. Engelmann, 1883), plate XIV.

page 15
right
© Wikimedia Commons, Benjah-bmm27 (PD)
page 20/21
From Paul Adam and Arnold Wyss, *Platonische und archimedische Körper, ihre Sternformen und polaren Gebilde*, 2nd ed. (Stuttgart: Verlag Freies Geistesleben, 1994), 66, 67, 76.

page 177
Illustration after Helmut Emde (redrawn and rearranged). From Helmut Emde, "Zur Geometrie räumlicher Strukturen," in *Diatomeen I: Schalen in Natur und Technik*, ed. Klaus Bach and Berthold Burkhardt (Stuttgart: Institut für leichte Flächentragwerke der Universität Stuttgart, 1984), 228.

All remaining illustrations and photographs
© Alexander Heinz

Note
The publisher has tried to identify any rights holders and to seek their permission in using copyrighted material. The publisher apologizes for any mistakes or omissions; in case of the presentation of relevant evidence, the references in question will of course be corrected in any future editions.

Acknowledgments

This book could not have become a reality without Haupt Verlag (for the German edition). In particular, I would like to thank Ms. Heidi Müller for her professional support and her lively enthusiasm for the project. I am very grateful to Frank Georgy for his dignified yet light and airy layout, his untiring dedication to any question of design, and his help in producing the illustrations. The photography team of the division of art at TU Dortmund deserves my thanks for all their advice and support that enabled me to take professional pictures of the models. I am grateful to the academic staff and fellow students in sculpture in the division of art for fruitful discussions of the relationship among art, science, and craft around the models.

I thank Peter N. Schiffer of Schiffer Publishing for his warm interest in my work, and for opening the door to the large English-speaking part of the world for my book and its content. My thanks are also due to the considerate translation from German into English by Katrin Binder, assisted by Neil Franklin and Alan Stott.

To the long-standing collaboration with my colleagues of the Austrian Fachverband der Geometrie (ADG), I owe thanks for their valuable suggestions, especially in the context of the annual Strobl seminar, with its intensive conversations and the space it provided to me, in which this project was able to develop freely. The same must be said of the collaboration with the TU Graz and the University of Innsbruck. My particular thanks go to Georg Glaser (University of Applied Arts Vienna) for his preface. For several helpful suggestions and support, I thank Georg Fuchs (Vienna), Rudi Neuwirt (Graz), Friedhelm Kürpig (Kornelimünster), Anuschka Pauluhn (Paul Scherrer Institute, Villingen), and especially Günter Maresch (University of Salzburg).

Meetings with lecturers and subject teachers of mathematics always were an enriching experience for me, especially at the University of Karlsruhe (now KIT), TU Dresden and TU Munich, University of Freiburg/Breisgau, and PH Fribourg/CH. At the latter, I especially thank Benedikt Finger and Yves Schubnel. A special thanks goes to my family and my friends. Without their well-meaning understanding, this project could not have flourished. Michael Doman (Kampen/Sylt) has supported my endeavors with unbroken enthusiasm over many years and contributed valuable suggestions and lively exchange of ideas regarding the geometry of polyhedra. My heartfelt gratitude for this! My thanks also to Gert Hansen (Copenhagen), Ueli Wittorf (Zurich), Fred Voss (Hannover), Jürgen Blasberg (Hagen), Michael von der Lohe (Hattingen), and Ernst Lehr† (Erdmannhausen) for their interested conversations and other helpful meetings. Finally, I am very grateful that my teachers at the Rudolf-Steiner-Schule Dortmund—both those who are still alive and those who have died—conveyed the first, rich basics in geometry to me during my years at school. I also thank my colleagues at the same school who have accompanied the coming into being of this book with lively interest.

Alexander Heinz

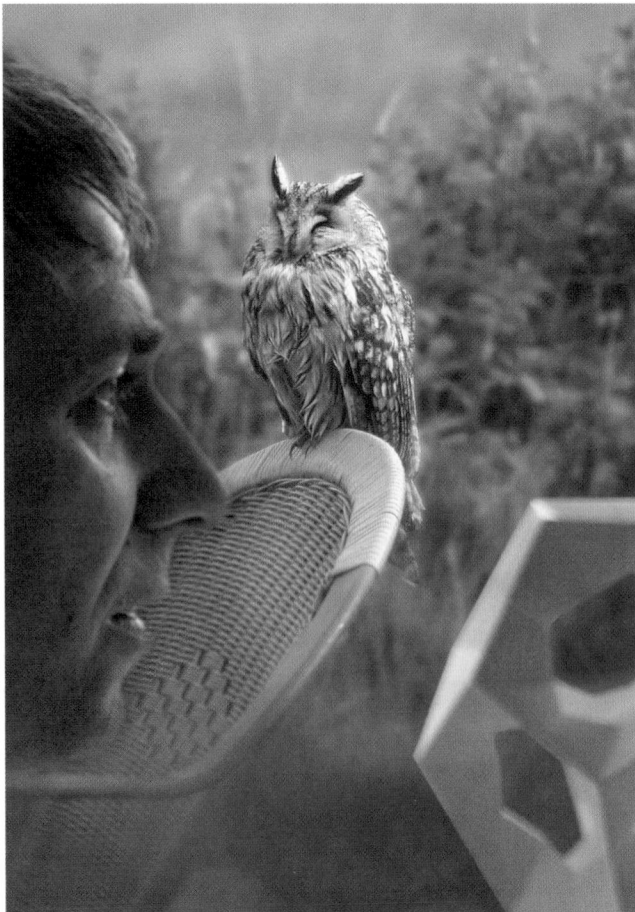

Alexander Heinz (b. 1968). Master's in bookbinding studies in art (for teaching in schools), at TU Dortmund. He teaches bookbinding and geometry at a school in Germany, leads workshops for young people and adults, and gives presentations at universities and other institutes of higher education in Germany, Austria, and Switzerland. Alexander Heinz has authored numerous articles about polyhedra forms and other subjects in geometry. With his free model projects, he wants to unite craft, art, and geometry.

More information at:
www.geomenta.com (in German) •